Handbook of Alloys

Handbook of Alloys

Lilly Armstrong

Willford Press
www.willfordpress.com

Published by Willford Press,
118-35 Queens Blvd., Suite 400,
Forest Hills, NY 11375, USA

ISBN: 978-1-64728-345-2

Cataloging-in-Publication Data

Handbook of alloys / Lilly Armstrong.
p. cm.
Includes bibliographical references and index.
ISBN 978-1-64728-345-2
1. Alloys. 2. Metallic composites. 3. Metals. I. Armstrong, Lilly.
TA483 .H36 2022
620.16--dc23

For information on all Willford Press publications
visit our website at www.willfordpress.com

Willford Press

Contents

Preface

The combination of a metal with another metal or nonmetal is known as an alloy. The resulting substance has properties which are often different from the original materials. The new material might have increased strength or toughness while still retaining all the characteristics of metals such as electrical conductivity, ductility and opaqueness. A defining feature of alloys is their metallic bonding character. Based on the arrangement of atoms in an alloy, they are usually classified into two types- substitutional alloys and interstitial alloys. Brass and bronze are examples of substitutional alloys while steel is a form of interstitial alloy. Stainless steel is an alloy which is a combination of substitutional and interstitial alloys. Some of the diverse topics covered in this book address the varied branches that fall under this category. The extensive content of this book provides the readers with a thorough understanding of the subject. It is meant for students who are looking for an elaborate reference text on alloys.

Given below is the chapter wise description of the book:

Chapter 1- The alloys which have a high strength to weight ratio and low density are called light alloys. Some of the metals which form lightweight alloys are aluminium, magnesium, titanium and beryllium. The topics elaborated in this chapter will help in gaining a better perspective about these light alloys, their properties and the hazards associated with them.

Chapter 2- Iron is the second most abundant metal in the earth's crust and is a trace element for all forms of life. Its alloys include steel, hematite, magnetite, etc. They are used in construction of strong structures of buildings and automobiles, manufacturing of magnets and magnetic storage media for computers, etc. This chapter has been carefully written to provide an easy understanding of the aspects related to iron and steel alloys.

Chapter 3- Aluminium is a silvery-white, soft, non-magnetic and ductile metal which belongs to the boron group. Its alloys are known for their improved strength, electrical conductivity and resistance to corrosion. All the diverse properties and applications of aluminium alloys have been carefully analyzed in this chapter.

Chapter 4- Magnesium is a chemical element which exists as a shiny gray solid. Its alloys are highly regarded for their combined properties of lightness and strength. It is used to make aircraft components, ladders, portable power tools and automotive wheels. This chapter discusses the diverse properties and applications of magnesium alloys in detail.

Chapter 5- Titanium is a lustrous, silver colored transition metal which has a low density and high strength. Beryllium is an alkaline earth metal which is strong, brittle and lightweight. It is used to form alloys along with nickel, iron, copper and other intermetallic compounds. The topics elaborated in this chapter will help in gaining a better perspective about the key properties and diverse uses of titanium and beryllium alloys.

Indeed, my job was extremely crucial and challenging as I had to ensure that every chapter is informative and structured in a student-friendly manner. I am thankful for the support provided by my family and colleagues during the completion of this book.

Lilly Armstrong

1

Introduction to Alloy

The alloys which have a high strength to weight ratio and low density are called light alloys. Some of the metals which form lightweight alloys are aluminium, magnesium, titanium and beryllium. The topics elaborated in this chapter will help in gaining a better perspective about these light alloys, their properties and the hazards associated with them.

Alloy

Alloy is a metallic substance composed of two or more elements, as either a compound or a solution. The components of alloys are ordinarily themselves metals, though carbon, a nonmetal, is an essential constituent of steel.

Alloys are usually produced by melting the mixture of ingredients. The value of alloys was discovered in very ancient times; brass (copper and zinc) and bronze (copper and tin) were especially important. Today, the most important are the alloy steels, broadly defined as steels containing significant amounts of elements other than iron and carbon. The principal alloying elements for steel are chromium, nickel, manganese, molybdenum, silicon, tungsten, vanadium, and boron. Alloy steels have a wide range of special properties, such as hardness, toughness, corrosion resistance, magnetizability, and ductility. Nonferrous alloys, mainly copper–nickel, bronze, and aluminum alloys, are much used in coinage. The distinction between an alloying metal and an impurity is sometimes subtle; in aluminum, for example, silicon may be considered an impurity or a valuable component, depending on the application, because silicon adds strength though it reduces corrosion resistance.

The term fusible metals, or fusible alloys, denotes a group of alloys that have melting points below that of tin (232 °C, 449 °F). Most of these substances are mixtures of metals that by themselves have low melting points, such as tin, bismuth, and lead. Fusible alloys are used as solder, in safety sprinklers that automatically spray out water when the heat of a fire melts the alloy, and in fuses for interrupting an electrical circuit when the current becomes excessive.

Many fusible alloys are formulated to melt at 90–100 °C (194–212 °F); for example, Darcet's alloy (50 parts bismuth, 25 lead, 25 tin) melts at 98 °C. By replacing half the tin in Darcet's alloy with cadmium, the alloy Wood's metal, which melts at 70 °C, is obtained.

Light Alloys and Metals

Light alloys and light metals have low density and high strength-to-weight ratios. They are generally characterized by low toxicity in comparison to heavy metals, although beryllium is an exception.

Light weight metals include aluminum, magnesium, titanium, and beryllium alloys. Aluminum and aluminum alloys are lightweight, non-ferrous metals with good corrosion resistance, ductility, and strength. Aluminum is relatively easy to fabricate by forming, machining, or welding. This metal is a good electrical and thermal conductor. Aluminum is also useful as an alloying element in steel and titanium alloys. Aluminum alloys are versatile metals with applications in almost every industrial and commercial segment.

Magnesium and magnesium alloys are non-ferrous metals with low density (relatively high strength to weight ratio), good ductility, moderate strength, and good corrosion resistance. Magnesium and magnesium alloys are used in a variety of industries as well as in aircraft, marine, and power tool applications.

Titanium and titanium alloys are non-ferrous metals with excellent corrosion resistance, good fatigue properties, and a high strength-to-weight ratio. Titanium's properties result in the use of titanium and titanium alloys in aircraft or airframe parts, jet engine super-alloy components, corrosion resistant chemical process equipment (valves, piping, and pumps), prostheses or medical devices, and marine equipment.

Beryllium has the second lowest density compared to the common structural light metal alloys (Al, Mg, and Ti). Magnesium has a density of 1.74 g/cc and aluminum has a density of 2.7 g/cc. Beryllium's unusually high Young's modulus (287 GPa) is a useful property in structural applications. The high modulus and low density make beryllium alloys useful for aerospace applications.

Beryllium has superior specific heat dissipation (heat dissipation per unit mass) compared to other metals because of beryllium's high specific heat and thermal conductivity. Beryllium also has a low thermal expansion coefficient (CTE). Many heat sinks and other thermal management products are fabricated from beryllium, beryllium alloys, and beryllium oxide ceramics due to beryllium's excellent thermal properties.

Properties and Production of Light Metals

Lightweight materials have become increasingly critical in the transportation manufacturing sectors, including aircraft, automobile, heavy truck, rail, ship, and defense manufacturing industries.

Light metal and alloys possess high strength-to-weight ratios and low density, and are generally defined by low toxicity as opposed to heavy metals, except for beryllium. Lightweight metals consist of aluminum, beryllium, titanium, and magnesium alloys.

Light metals are often used for materials and operations where lightweight and improved performance properties are required. Common applications include chemical process, marine, aerospace, and medical applications.

Lighter vehicles that are designed for consumers, as well as the industry and military sectors, consume less fuel and provide a better performance. In addition to carrying larger loads, lighter vehicles can travel the same distances at reduced cost and release less carbon dioxide.

Properties of Lightweight Metals

Aluminum and aluminum alloys are non-ferrous metals that are lightweight and characterized by excellent strength, ductility, and corrosion resistance. Aluminum is a good conductor of heat and electricity and is also used as an alloying element in titanium/steel alloys. Aluminum alloys are also resourceful.

Aluminum is lighter than titanium but not as strong. Aluminum alloys have excellent rigidity and flexibility that increase with decreasing temperature.

Below zero temperature, most of the aluminum alloys display a slight change in properties such as impact strength and tensile strengths. Yield may increase and elongation may slightly decrease. Aluminum alloys do not have a ductile-to-brittle transition.

Similar to aluminum and aluminum alloys, magnesium and magnesium alloys are non-ferrous metals that are characterized by moderate strength, good ductility, low density, and excellent corrosion resistance. The materials have tensile and other properties that rely on several factors such as the condition, composition, heat treatment, details of fabrication, etc.

Magnesium has no definite yield point, and magnesium alloys are susceptible to notches and other stress-raisers, which considerably reduce their endurance limits. Magnesium-base materials posses low modulus of elasticity, and have high unit resilience.

Unlike molten aluminum, molten magnesium does not react with tool steels and thus leads to longer die life. Also, its reduced heat input and low erosion reduce the tendency for thermal fatigue.

Titanium and titanium alloys have high strength-to-weight ratios, good fatigue properties, and excellent corrosion resistance. Titanium is also considered as a refractory metal, which means it is highly resistant to heat and wear. It does not corrode in chlorine or sea water, and it is one of the metals that are resistant to aqua regia.

Commercially pure titanium and titanium alloys are non-magnetic. The thermal conductivity of all titanium alloys is quite low for a metal. Temperature considerably affects the physical properties of titanium. The alloy grades, predominantly the high strength materials, can retain tensile and proof strengths up to relatively higher temperatures when compared to commercially pure grades.

Titanium alloys have excellent high cycle fatigue strengths as opposed to their tensile strengths. The toughness is based on texture, microstructure, strength, and composition, which are interrelated.

Properties of Beryllium and Beryllium Alloys

Beryllium is a useful alloying element. It is characterized by low density, high rigidity, high strength, structural stability, and reflectivity. Beryllium is distinguished among metals in terms of specific rigidity.

The metal is known to have the second lowest density compared to common structural light metal alloys such as magnesium, aluminum, titanium. Beryllium has high thermal conductivity and specific heat and hence, has excellent specific heat dissipation compared to other metals. The metal has an unusually high Young's modulus.

Beryllium's density is 30% less than that of aluminum, whereas its rigidity is 50% greater than that of steel. The metal's specific rigidity is about four times greater than that of composites and six times greater than that of other alloys or metals. It is resistant to corrosion in normal ambient conditions and at elevated temperatures.

Production of Lightweight Metals

Aluminum Alloys

Aluminum alloys can be developed by rolling, casting, forging, extrusion, and drawing; however varying compositions lend themselves to certain process more readily than others. The hot extrusion of aluminum metal is very important as it allows the production of almost any type of shape in cross section.

In this process, a preheated billet is mounted onto the press container and pressed via a shaped steel die with a ram. While most of the aluminum alloys can be extruded, the 6000 series alloys form the bulk of commercial extrusion, providing an optimum combination of speed and ease of extrusion together with the ability to form thin sections and complex shapes with good surface finish.

Although the extrusion process is versatile, it can provide only 2D shapes. In contrast, casting allows the production of complex 3D products. Over the past few years or so, the development work on the extrusion process has improved the ability of castings to be employed in significantly stressed applications.

Sand, die casting, and permanent mold methods are being used. The choice of the process relies on the quantity of castings needed, property, quality and property considerations, and end use. Earlier, die castings were not used for heat treatment, but this limitation has been overcome with advanced methods.

Production of Magnesium and Magnesium Alloys

Magnesium can be cast through a wide range of techniques such as sand casting, permanent mold casting, high-pressure diecasting, and squeeze and semi-solid casting. Diecasting is the most prevalent casting technique for magnesium.

In this procedure, thin-walled, complex components are created at high production rates supported by the low-heat-content per volume of molten metal. Cold chamber and hot chamber machines are used for magnesium.

To ensure optimum performance, higher shot speeds can be used for magnesium, particularly for thin-walled components. Vacuum diecasting is one of the diecasting process variants that can create parts with better properties and lower porosity compared to the traditional diecasting process.

Semi-solid casting methods can also be used to fabricate magnesium, but using magnesium alloy granules instead of liquid magnesium. Semi-solid molding is generally used for smaller components.

When assessing different processes and alloys for magnesium casting, several factored should be taken into account to ensure a high-quality component at a lower cost. These factors include the tooling costs, post-casting operations, and end-use applications.

Magnesium casting can achieve a die life of three to four times more that can be accomplished with aluminum. Magnesium is also known to be the easiest of structural metals to machine, and it is the standard of the cutting tool sector compared to machinability of other metals.

Production of Titanium and Titanium Alloys

Kroll process is the main production process used for titanium metal. In this procedure, the principle ore rutile is treated with chlorine gas to create titanium tetrachloride, which is subsequently purified and reduced to a metallic titanium sponge through reaction with sodium or magnesium.

This titanium sponge is eventually subjected to an alloying and melting process. However, this process is laborious, making it very expensive. A method was later developed to produce high-quality titanium alloy powders, which was effective for processing this metal into commercial products.

Several steps are involved in the conversion of purified titanium sponge to a form suited for structural purposes. Consolidation into titanium ingot is carried out in an argon or vacuum environment.

Alloying elements, sponge, or sometimes recycled scrap are mechanically compacted and welded into an extended, cylindrical electrode, which is later melted in a copper crucible by passing electricity through it. This ingot is again melted in a similar fashion to ensure even distribution of alloying elements.

Another consolidation process is cold-hearth melting, which is performed within a vacuum or argon chamber comprising of a copper crucible. Heating is achieved with helium/argon plasma torches or a multiple electron-beam, and the molten metal passes over the hearth lip and finally into a water-cooled copper mold.

The cold-hearth process is suitable for isolating high-density contaminants, which collect at the base of the hearth. Consolidated ingots are then processed into mill products such as plate, tubing, bar, wire, billet, etc. by steel facilities.

Production of Beryllium and Beryllium Alloys

High purity beryllium hydroxide serves as the input material for important applications of the element such as beryllia ceramics, copper beryllium alloys, and pure beryllium metal manufacturing. To create high-purity beryllium metal, the hydroxide form is first dissolved in ammonium bifluoride and then heated to over 900°C, yielding a molten beryllium fluoride.

After casting the beryllium fluoride into molds, it is combined with molten magnesium in crucibles and then heated. This process separates pure beryllium from the magnesium slag, forming beryllium spheres. Surplus amount of magnesium is then burned off by further treating it in a vacuum furnace, leaving behind 99.99% pure beryllium.

The beryllium spheres are usually changed to powder through isostatic pressing, producing a powder that can be employed in the production of pure beryllium metal shields or beryllium-aluminum alloys.

Processing of Lightweight Metals

Lightweight metals are processed in a variety of ways such as melt processing, powder processing, thermo-mechanical processing, forming, coatings, and joining and assembly.

Melt Processing

Metal casting, which involves pouring molten metal into a die or mold followed by cooling it to solidify the component, is an ancient process and even today offers great potential to remove weight off the metal structures. There are three advanced melt processing techniques: Thin-wall casting, high-integrity casting, and dissimilar-metal casting.

Thin-wall casting is a process where several types of metals, such as aluminum and steel can be cast. However, some complications occur when working with molten metals; such as maintaining proper flow and inhibiting the metal from solidifying prior to filling the mold.

In high-integrity casting, certain products should meet unique standards, without microstructures and porosity that are present throughout the cast part.

In dissimilar-metal casting, two or more metals are used in a single casting which provides considerable benefits. Here, a product can be cast so that parts of it are formed from one type of metal and other parts are formed from a different metal, employing the materials' various properties where they are most needed.

Powder Processing

In this process, heat, pressure, or both are applied to the powder to create hard products with internal mixtures that confer new, valuable properties to the products. The powder is produced by melting the metal and forcing through a nozzle while subjecting it to a water spray or inert gas. This breaks the metal into small droplets, which are eventually cooled down to form a powder.

Next, the metal powder is rolled, extruded, sprayed or pressed to produce blanks for additional shaping or to directly mold into shapes as almost finished components.

Thermo-mechanical Processing

Thermo-mechanical processing (TMP) improves the properties of metals, enables lightweight design, and lowers manufacturing costs. TMP involves the concurrent control and use of thermal and deformation processing to acquire components and materials that have improved properties and deliver better performance.

TMP provides controls at each process stage. The shaping time and temperature are simultaneously controlled, together with the amount of deformation at each step of the process in addition to controlling the cooling. This process of controlled heating, forming, and cooling at each step of the process enables engineers to exploit material microstructures in many innovative ways.

TMP provides better control over the end product, potentially removing the need for extra reheating and quenching steps and making the production process less costly and more efficient. TMP can create new, improved properties for aluminum alloys, titanium alloys, and other metals. The process improves materials' strength, resistance to fracture, and resistance to fatigue-induced degradation.

Forming

Forming involves a variety of methods that are used to convert a metal into a usable form. Forming processes consist of bending, extrusion bending, brake bending, roll bending, press drawing, etc.

Bending is often used to form magnesium sheet and extrusions, and the bend radius is a major factor in the bending operation.

Rubber pad forming is most conventional forming process, where a rubber pad is used to replace metal dies during metal forming; hydroforming and marforming are a variation of this process.

Impact extrusion, also called back extrusion/reverse extrusion, has been traditionally used on magnesium alloys and can be used to acquire perfectly symmetrical tubular parts; it can be employed to make components that cannot be made by other techniques.

Spinning involves the use of a spinning tool to produce a component in line with the shape of a mandrel or form block in a lathe; this process is the most economical way for forming products where only a few parts are required. Similar to the spinning process, flow forming is formed in line with the shape of the form block without altering the section thickness of the blank material.

Forging is performed at high temperatures for all types of metals; in this process, metals are forged by pounding them with a hammer from various directions until the required shape is obtained.

Joining and Assembly

Methods to join lightweight materials, predominantly metals, are becoming more and more important in the development of hybrid structures for engineering applications. Suitable joining techniques are required to consistently join these materials.

Some of the joining methods include adhesive bonding, mechanical fastening, and welding. The processes can be used either separately or in combination in a single method to ensure a durable joint surface. The joining methods come with their own advantages and disadvantages, and the most optimum technique will rely on the application.

Adhesive bonding is a solid state joining method that depends on the formation of intermolecular forces between the workpieces and the polymeric adhesive for the formation of joints. Mechanical fastening uses extra clamping components without combining the joint surfaces.

Traditional welding processes such as gas metal arc welding, gas tungsten arc welding, submerged arc welding, and shielded metal arc welding have been employed to weld dissimilar materials in metal-to-metal joints. However, the high energy inputs of these processes lead to material metallurgical mismatch.

New emerging methods, such as ultrasonic welding, laser welding, friction stir welding, and friction spot welding, are considered to be more viable. Ultrasonic welding is a solid state joining method characterized by low energy input.

In this method, coalescence is initiated by applying localized high-frequency vibration energy with a modest clamping force. Laser welding complements the processing and fabrication of joints, which are difficult to obtain with other welding techniques.

Friction spot joining is similar to linear friction stir welding, but there is no linear movement of the tool during the time of friction spot joining. This process includes three phases such as plunging, stirring, and retracting. Coating is yet another method used for processing lightweight metals.

Hazards Associated with Light Metals

Light metals, most commonly aluminum, magnesium and titanium, and alloys containing them, are used in many industrial applications where lightness, hardness and ductility and resistance to corrosion are needed. They are widely considered to be near incapable of creating a frictional spark which is hot enough to ignite a flammable atmosphere. However, there are two exceptions to this where light metal alloys can be potential ignition sources in flammable atmospheres (e.g. methane): (a) thermite reaction, and (b) incendive chips (titanium).

Potentially hazardous light alloys are those in which the total weight of aluminum, magnesium and titanium together exceeds 15%, and/or in which the content of magnesium and titanium together exceeds 6% by weight.

Thermite Reaction

Light metals and their alloys have an affinity for oxygen and when they are brought into close contact with oxygen bearing material, such as iron oxide (rust) in the presence of moderate heat, then a chemical reaction occurs, called a thermite reaction.

- This thermite reaction can generate temperatures over 2000/C.
- The heat necessary to start the thermite reaction can be generated from frictional contact or from merely a glancing blow.
- A similar reaction can also be produced if either a smear of light metal is left on a rusty surface and is then struck with a glancing blow or where a rusty metal object strikes a surface coated with aluminum paint.
- Aluminum foils, widely used for cooking and wrapping candies and cigarettes will also produce a thermite reaction if laid over rusty metal and struck a glancing blow.
- Aluminum pop cans and lunch boxes can also potentially produce a thermite reaction if they come into contact with rusty steel.

Incendive Chips

Titanium and its alloys can produce highly incendive burning chips when they are struck or rubbed on various hard materials. These 'sparks' are easily capable of igniting a flammable atmosphere.

2

Iron and Steel Alloys

Iron is the second most abundant metal in the earth's crust and is a trace element for all forms of life. Its alloys include steel, hematite, magnetite, etc. They are used in construction of strong structures of buildings and automobiles, manufacturing of magnets and magnetic storage media for computers, etc. This chapter has been carefully written to provide an easy understanding of the aspects related to iron and steel alloys.

Iron

Iron is the second most abundant metal (after aluminum) in the Earth's crust and the most abundant element (in terms of mass) in the Earth as a whole. Its predominance in the Earth's core helps produce our planet's magnetic field, which in turn shields the Earth's surface by deflecting charged particles coming from the solar wind. Iron and nickel are notable for being the final elements produced by nuclear fusion reactions within stars—processes that are collectively known as stellar nucleosynthesis. Scientists think that iron (and other elements) on Earth were once part of the material ejected by a supernova (exploding star).

Iron is a necessary trace element for all known forms of life. Iron-containing enzymes help transport various soluble gases (including oxygen and carbon dioxide) and catalyze oxidation reactions in living organisms. In addition, given its high abundance, low cost, and high strength, iron is the most widely used of all metals and constitutes about 95 percent of the tonnage of all metals produced worldwide. Its alloys, particularly steels, are used to make large, strong structures such as buildings, automobiles, and the hulls of ships. Iron is also used for making magnets, and iron(III) oxide in particular is used for the manufacture of magnetic storage media for computers. Iron and its alloys, however, tend to rust and need to be protected by such methods as painting, galvanization, or the application of plastic coatings.

Production of Iron from Iron Ore

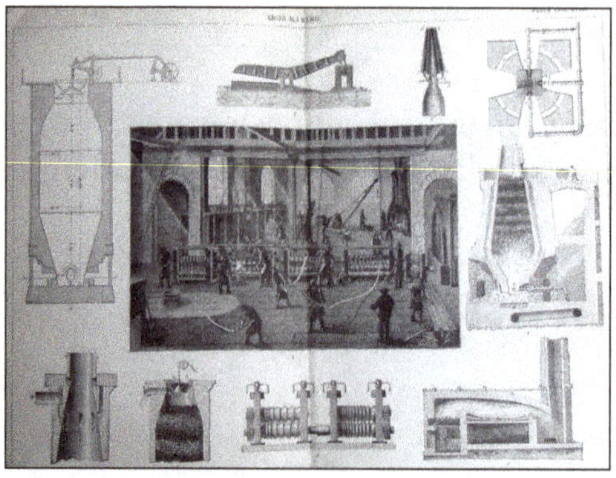

How iron was extracted in the nineteenth century.

Iron production constitutes about 95 percent of the tonnage of all metals produced worldwide. Industrially, it is produced starting from iron ores, principally hematite (nominally Fe_2O_3) and magnetite (Fe_3O_4) by reducing the ore with carbon in a blast furnace at temperatures of about 2,000 °C. Iron ore, carbon (in the form of coke), and a *flux* such as limestone are fed into the top of the furnace, while a blast of heated air is forced into the furnace from the bottom.

In the furnace, the coke reacts with oxygen in the air blast to produce carbon monoxide:

$$6\,C + 3\,O_2 \rightarrow 6\,CO$$

The carbon monoxide reduces the iron ore (represented as hematite in the chemical equation below) to molten iron, becoming carbon dioxide in the process:

$$6\,CO + 2\,Fe_2O_3 \rightarrow 4\,Fe + 6\,CO_2$$

The flux is present to melt impurities in the ore, principally silicon dioxide sand and other silicates. Common fluxes include limestone (principally calcium carbonate, $CaCO_3$) and dolomite (magnesium carbonate). Other fluxes may be used, depending on the impurities that need to be removed from the ore. In the heat of the furnace, the limestone flux decomposes to calcium oxide (quicklime, CaO):

$$CaCO_3 \rightarrow CaO + CO_2$$

Then calcium oxide combines with silicon dioxide (SiO_2) to form calcium silicate *slag*:

$$CaO + SiO_2 \rightarrow CaSiO_3$$

The slag melts in the heat of the furnace, at a temperature at which silicon dioxide would not have melted. In the bottom of the furnace, the molten slag floats on top of the denser molten iron, and spouts in the side of the furnace may be opened to drain off either the iron or the slag. The iron, once cooled, is called pig iron. The slag can be used in road construction or to improve mineral-poor soils for agriculture. Pig iron is later reduced to steel, using convertors.

Iron ore is produced in 48 countries, of which the five largest producers are China, Brazil, Australia, Russia, and India, accounting for 70 percent of world iron ore production. Approximately 1,100 Mt (million tons) of iron ore was produced in the world in 2000, with a gross market value of approximately 25 billion US dollars. This iron ore was used to produce approximately 572 Mt of pig iron.

Notable Characteristics

Iron is a ductile, malleable, gray-colored metal and is a good conductor of heat and electricity. It is attracted by magnets and can be readily magnetized. In chemical terms, it is classified as a transition metal. It is located in period 4 of the periodic table, situated between manganese and cobalt. In addition, it lies at the top of group 8 (former group 8B). Iron, cobalt, and nickel have a number of similar properties and were once grouped together as group 8B.

Iron is believed to be the tenth most abundant chemical element in the universe. Making up about 5 percent of the Earth's crust, it is second in abundance to aluminum among metals and fourth in abundance among the elements. In addition, it makes up about 32 percent of the mass of the Earth as a whole, thus being the most abundant element (by mass) on the planet. The concentration of iron in the various layers of the Earth ranges from very high at the inner core to only a few percent in the outer crust.

Iron is extracted from iron ore and is almost never found in the free elemental state. To obtain elemental iron, the impurities must be removed by chemical reduction. Iron is used in the production of steel, an alloy of different metals and some nonmetals, particularly carbon. (The many iron-carbon allotropes, which have very different properties).

The universally most abundant of the highly stable nuclides is ^{56}Fe. It is formed by nuclear fusion reactions in stars. In addition, when a very large star contracts at the end of its life, internal pressure and temperature rise, allowing the star to produce progressively heavier elements, although these are less stable than the elements around mass number 60, known as the "iron group". This process leads to a supernova.

Isotopes

Naturally occurring iron consists of four isotopes: 5.85 percent of slightly radioactive ^{54}Fe (half-life $>3.1\times10^{22}$ years), 91.75 percent of stable ^{56}Fe, 2.12 percent of stable ^{57}Fe, and 0.28 percent of stable ^{58}Fe. In addition, it appears that the naturally occurring radioactive isotope ^{60}Fe, with a half-life of 1.5 million years, is now extinct, but it can be produced synthetically. Much of the past work on measuring the isotopic composition of iron centered on determining ^{60}Fe variations due to processes accompanying nucleosynthesis (that is, through meteorite studies) and ore formation.

The isotope ^{56}Fe is of particular interest to nuclear scientists. A common misconception is that this isotope represents the most stable nucleus possible, and that it would be impossible to perform fission or fusion on ^{56}Fe and still liberate energy. This is not the case, as both ^{62}Ni and ^{58}Fe are more stable.

In phases of the meteorites *Semarkona* and *Chervony Kut,* a correlation between the concentration of ^{60Ni}, the daughter product of ^{60}Fe, and the abundance of the stable iron isotopes could be found, suggesting the existence of ^{60}Fe at the time of formation of the solar system. The abundance of ^{60}Ni present in extraterrestrial material may provide further insights into the origin and early history of the Solar System.

Compounds of Iron

- Iron(II) oxide or ferrous oxide (FeO): This black-colored powder needs to be handled with care as it readily ignites and can cause explosions. It has a mineral form known as wüstite. It is an example of a non-stoichiometric compound, as the ratio of iron and oxygen atoms can vary due to vacancies in the crystal structure. It is a pigment and is used in some tattoo inks. It has been approved by the U.S. Food and Drug Administration (FDA) for use in cosmetics.

- Iron(III) oxide:, ferric oxide, or red iron oxide (Fe_2O_3): This compound corresponds to iron rust. Its mineral form, known as *hematite,* is mined as the main ore of iron and is used in the production of iron in a blast furnace. It has ferromagnetic properties and is therefore useful for magnetic storage media in computers. A fine powder of ferric oxide is known as "jeweler's rouge", "red rouge," or simply "rouge". It is used to put the final polish on metallic jewelry and lenses. It is also a pigment in products known as "Pigment Red 101" and "Pigment Brown 6," which have been approved by the FDA for use in cosmetics.

- Iron (II, III) oxide or ferrous ferric oxide (Fe_3O_4): This oxide is found in nature as the black-colored mineral *magnetite* or *lodestone,* and it is a major ore of iron. It is also seen on Mars. This oxide of iron tends to occur when iron corrodes underwater, and so is often found inside tanks or below the waterline of ships.
- Iron(III) oxide-hydroxide, hydrated iron oxide, or yellow iron oxide (FeO(OH)): This solid material (in anhydrous or hydrated forms) has colors ranging from yellow through dark brown to black. It occurs naturally as minerals goethite, siderogel, feroxyhyte, limonite, and lepidocrocite. It is used in aquarium water treatment as a phosphate binder. The FDA has approved its use, in the form of "Pigment Yellow 42," in cosmetics and some tattoo inks.
- Iron(III) chloride or ferric chloride ($FeCl_3$): This crystalline compound appears dark green by reflected light, but purple-red by transmitted light. It is acidic, toxic, and highly corrosive. It absorbs moisture from the air, producing fumes of hydrochloric acid (HCl). When dissolved in water, it undergoes hydrolysis and gives off a great deal of heat. The resulting brown, acidic solution, which is corrosive, is used as a coagulant in treating sewage and drinking water and as an etchant for copper-based metals (such as those found in printed circuit boards) and stainless steel. It is used (in the anhydrous form) as a catalyst for certain organic synthetic reactions.
- Iron(II) sulfate or ferrous sulfate ($FeSO_4$): This ionic compound can be found in various states of hydration (that is, a varying number of water molecules can be attached to each molecule of ferrous sulfate). The heptahydrate ($FeSO_4{\cdot}7H_2O$) is also called green vitriol or copperas. It is used in the manufacture of inks, in wool dyeing as a mordant (agent that sets the dye), and in water purification as a substitute for aluminum sulfate. In horticulture, it is used as a lawn conditioner and moss killer. It is also an additive found in various foods, and it is used to treat iron-deficiency anemia.

Applications

Of all metals, iron makes up the highest level of production and usage. Its combination of low cost and high strength makes it indispensable, especially for applications such as automobiles, hulls of large ships, and structural components for buildings. Some of the various alloys of iron and their applications are given below.

- Pig iron: It has a carbon content of 4–5%, and varying amounts of contaminants such as sulfur, silicon, and phosphorus. Its only significance is that it forms an intermediate in the steps leading from iron ore to cast iron and steel.
- Cast iron: It contains 2–4.0 percent carbon, 1–6 percent silicon, and small amounts of manganese. Contaminants present in pig iron (such as sulfur and phosphorus) that negatively affect the material properties have been reduced to an acceptable level. Cast iron has a melting point in the range of 1420–1470 K, which is lower than that of either of its two main components. Consequently, it is the first product to be melted when carbon and

iron are heated together. Its mechanical properties vary greatly, depending on the form of carbon in the alloy. For instance, "white" cast irons contain carbon in the form of cementite, or iron carbide. This hard, brittle compound dominates the mechanical properties of the white cast irons, rendering them hard but unresistant to shock. The broken surface of a white cast iron is full of fine facets of the broken carbide—a pale, silvery, shiny material, hence the appellation. In gray iron, the carbon exists free as fine flakes of graphite, and also renders the material brittle due to the stress-raising nature of the sharp-edged flakes of graphite. A newer variant of gray iron, referred to as ductile iron, is specially treated with trace amounts of magnesium to alter the shape of graphite to sheroids (or nodules), vastly increasing the material's toughness and strength.

- Carbon steel: It contains 0.4–1.5 percent carbon, with small amounts of manganese, sulfur, phosphorus, and silicon.
- Wrought iron: It contains less than 0.2 percent carbon and is a tough, malleable product, not as fusible as pig iron. If honed to an edge, it loses it quickly. Wrought iron is characterized, especially in old samples, by the presence of fine "stringers" (or filaments) of slag entrapped in the metal. Wrought iron resists rusting when used outdoors. It has, however, largely been replaced by mild steel for "wrought iron" gates and blacksmithing. Mild steel does not have the same corrosion resistance but is cheaper and more widely available.
- Alloy steels: They contain varying amounts of carbon and different metals such as chromium, vanadium, molybdenum, nickel, and tungsten. They are used for structural purposes, as their alloy content raises their cost and necessitates justification of their use. Recent developments in ferrous metallurgy have produced a growing range of microalloyed steels, also termed "HSLA" (high-strength, low-alloy) steels, containing tiny additions to produce high strengths and often spectacular toughness at minimal cost.

The main drawback in using iron and its alloys is that they suffer from rust if not protected in some way. Painting, galvanization, plastic coating, and bluing are some techniques used to protect iron from rust by excluding water and oxygen, or by sacrificial protection.

Iron in Organic Chemical Reactions

The use of iron metal filings in organic syntheses is mainly for the reduction of nitro compounds. Additionally, iron has been used for desulfurization reactions, reduction of aldehydes, and the deoxygenation of amine oxides.

Iron in Biology

Iron is an essential element for all known organisms. In most cases, it is stably incorporated within proteins that belong to the class known as metalloproteins. By contrast, in the exposed ("free" form), it causes production of free radicals that are generally toxic to cells. To say that iron is free does not mean that it is free floating in bodily fluids. It binds avidly to virtually all biomolecules, so it will adhere nonspecifically to cell membranes, nucleic acids, proteins, and so forth.

In many animal species, iron (in the form of the ferrous ion, Fe^{2+}) is present in the heme complex, an essential component of proteins such as hemoglobin (involved in carrying oxygen) and cytochromes (involved in cellular respiration and other processes). Inorganic iron involved in redox reactions is also found in the iron-sulfur clusters of many enzymes, such as nitrogenase (involved in the synthesis of ammonia from nitrogen and hydrogen) and hydrogenase.

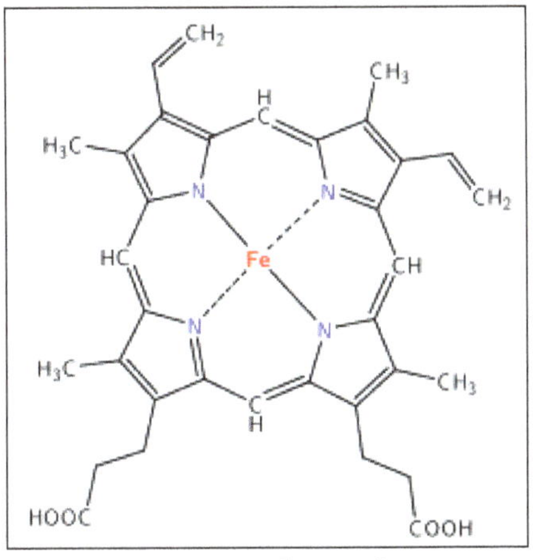

Structure of Heme b.

In addition, a number of non-heme iron proteins are enzymes that are responsible for a wide range of functions in several life forms. When the body is fighting a bacterial infection, the body sequesters iron inside of cells (mostly in the storage molecule ferritin), so that it cannot be used by bacteria.

Iron distribution is heavily regulated in mammals, both as a defense against bacterial infection and because of the potential biological toxicity of iron (noted below). The iron absorbed from the duodenum binds to transferrin and is carried by blood to different cells. There it gets incorporated into target proteins.

Nutrition and Dietary Sources

Good sources of dietary iron include meat, fish, poultry, lentils, beans, leaf vegetables, tofu, chickpeas, black-eyed peas, strawberries, and farina.

Iron provided by dietary supplements is often found as iron(II) fumarate. Iron sulfate is as well absorbed, and less expensive. The most bioavailable form of iron supplement (ten to fifteen times more bioavailable than any other) is iron amino acid chelate. The Recommended Dietary Allowance (RDA) for iron varies considerably based on the age, gender, and source of dietary iron (heme-based iron has higher bioavailability).

Precautions

Excessive iron is toxic to humans because excess ferrous iron reacts with peroxides in the body, producing free radicals. Iron becomes toxic when it exceeds the amount of transferrin present to bind free iron. In excess, uncontrollable quantities of free radicals are produced.

Iron uptake is tightly regulated by the human body, which has no physiologic means of excreting iron and regulates iron solely by regulating uptake. Too much ingested iron can damage the cells of the gastrointestinal tract directly and may enter the bloodstream by damaging the cells that would otherwise regulate its entry. Once in the bloodstream, it can damage the cells of organs such as the heart and liver. This could potentially lead to long-term organ damage and even death from overdose.

Humans experience iron toxicity above 20 milligrams (mg) of iron for every kilogram (kg) of weight, and 60 mg/kg is a lethal dose. Over-consumption of iron, often the result of children consuming large quantities of ferrous sulfate tablets intended for adults, is the most common toxicological cause of death in children under six. The Dietary Reference Intake (DRI) lists the Tolerable Upper Intake Level (UL) for adults as 45 mg/day. For children under fourteen years old, the UL is 40 mg/day.

If iron intake is excessive, it may lead to iron overload disorders such as hemochromatosis. Iron overload disorders are based on a genetic inability to regulate iron uptake. Many people have a genetic susceptibility to iron overload without realizing it and without knowing a family history of the problem. For this reason, people should not take iron supplements unless they suffer from iron deficiency and have consulted a doctor. Blood donors are at special risk of low iron levels and are often advised to supplement their iron intake.

The medical management of iron toxicity is complex. One element of the medical approach is a specific agent called deferoxamine, used to bind and expel excess iron from the body in case of iron toxicity.

Iron as a Metaphor

The word "iron" is used metaphorically as an adjective, to describe certain iron-like traits, including sturdy, strong, stern, harsh, unyielding, inflexible, rigid, or robust. We are now familiar with many materials that are stronger and sturdier than iron, such as diamond or steel. Still, the "iron" metaphor persists among many cultures, apparently rooted in the Iron Age, when iron defeated bronze.

Iron Alloy

Iron alloy, as the name suggests, is an alloy that have iron as the principal component. Iron is used as a constituent in most of the commercial alloys. For example, iron is the major component of wrought and cast iron and wrought and cast steel. Iron can be alloyed with manganese, silicon, vanadium, chromium, molybdenum, niobium, selenium, titanium, phosphorus, or other elements for commercial use. Some iron alloys are also used as addition agents in steel-making. There are special-purpose iron alloys that demonstrate exceptional characteristics, like magnetic properties, electrical resistance, thermal expansion, corrosion resistance, and heat resistance, etc.

Iron Alloy Castings

Iron castings are produced by molding iron alloys or molten iron. Gray iron is possibly one of the oldest worked metal used for casting. The metal is also one of the most abundant and least expensive material.

Gray iron was the original cast iron, though it has recently been replaced in various applications by other iron-carbon alloys with higher tensile strength. Ductile iron, as the name suggests, has high ductility than traditional iron materials, such as gray iron. These materials incline to be brittle and are prone to fracture under high tensile stress.

Iron is cast as like any other metal. It is poured into a mold and extracted after cools down. There are different types of casting methods, through which iron can be molded, which include shell molding, green sand molding, and centrifugal molding. The procedure through which the iron is caste can effect the mechanical properties of the metal, especially in regard to its cooling rate. Iron castings are used in a variety of applications, including - automotive, appliance, agricultural, and machinery industries. Cast iron components are used pump housings, engine blocks, electrical boxes, decorative castings, and more.

Cast Iron

The term Cast Iron describes a large family of ferrous alloy. It contains 2-4% carbon, along with silicon, manganese. Cast irons are produced in specified shapes for processing by machining, heat treating, or assembly. Sometimes, to meet certain requirement it can be forged or rolled. Cast iron is produced by melting pig iron and then combining it with steel or scrap iron.

Normally, cast iron contains silicon up to 3%. Though, it may vary while making special compositions, such as:

- Silal: Silicon up to 6%.
- Duriron: Silicon up to 12%.

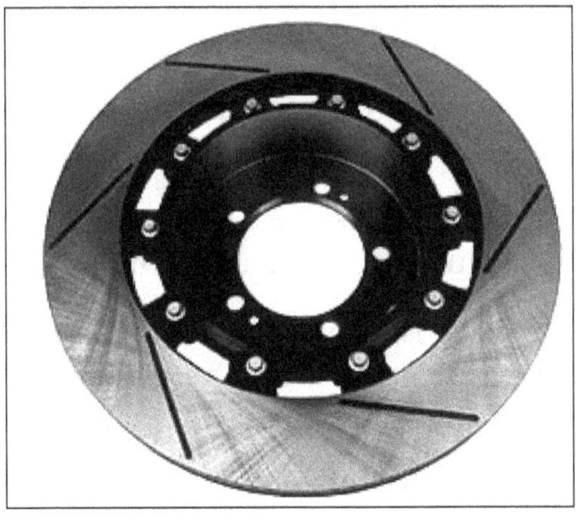

Cast iron is produced by melting pig iron. Often, this melting is done along with scrap steel and scrap iron. It also involves a number of steps that leads to the removal of undesirable elements (phosphorus and sulfur).

- The main purpose of this processing is to reduce carbon and silicon content to the desired levels. Then, the addition of other elements to the melt takes place.

- A small blast furnace known as a cupola is used to melt iron.
- After melting is complete, the molten iron is removed from the forehearth of the blast furnace.
- The final form is made by casting.

Types of Cast Irons

During solidification, the major proportion of the carbon precipitates in the form of graphite or cementite. When solidification is just complete, the precipitated phase is embedded in a matrix of austenite which has an equilibrium carbon concentration of about 2 wt%. On further cooling, the carbon concentration of the austenite decreases as more cementite or graphite precipitates from solid solution. For conventional cast irons, the austenite then decomposes into pearlite at the eutectoid temperature. However, in grey cast irons, if the cooling rate through the eutectoid temperature is sufficiently slow, then a completely ferritic matrix is obtained with the excess carbon being deposited on the already existing graphite.

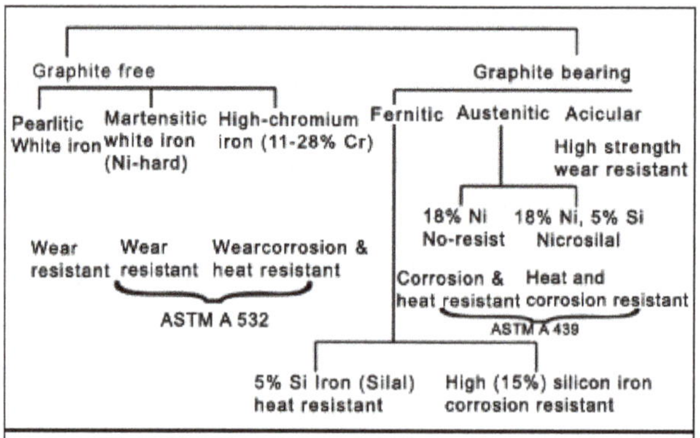

Classification of special high - alloy cast iron

- White Cast Iron: White cast iron is hard to be machined as it's tough and brittle.
- Grey Cast Iron: Due to the presence of microstructure of graphite in transformed-austenite and cementite matrix, grey cast irons are softer.
- Spheroidal Graphite Cast Iron: The chemical composition of the cast iron is similar to that of the grey cast iron but with 0.05 wt% of magnesium.
- Ductile Cast Iron: One of the most popular alloys used in casting. It has wide applications including automobile components, industrial machinery, wind turbine electrical energy generation, valves, air conditioning machinery, lawn and garden equipment and agricultural products.
- Spheroidal Graphite Cast Iron: Spheroidal graphite cast iron is known for its excellent toughness and has wide applications, e.g. crankshafts.

Cast Iron Alloy Castings

Cast iron alloys are widely used in casting applications for their characteristics, such as strength, good strength to weigh ratio, economical price, and availability in abundance, capability to produce complex geometries. The cast iron castings are used in a variety of applications, such as automotive, agriculture, appliance, and more.

Advantages in Casting

- A family of materials that is capable to meet a variety of engineering and manufacturing applications (the "family", includes - gray iron, ductile iron, compacted graphite iron, malleable iron, and white iron).
- Capability of casting with inserts of other materials.
- Capability to produce highly complex geometries and sections in variety of sizes.
- Possibility to cast intricate shapes as well as very thin to very thick section sizes.
- Good strength to weight ratio.
- Generally economical than competing materials and relatively low cost per unit of strength than other materials.
- Superior damping capacity, especially in Gray Irons.
- Capability of redesigning and combining two or more parts from other materials into a single casting, thereby reducing assembly time and cost.
- Different types of casting processes for low, medium or high production.
- Variety of iron castings can be used without heat treatment (as cast) however, if required, they can be heat treated to improve the overall properties or specific properties such as surface hardness.

Alloy K

Alloy K is an iron-nickel-cobalt alloy designed for precise and uniform thermal expansion chracteristics. The coefficient of thermal expansion of nickel and iron is more or less the same. However, adding nickel to iron can result in an alloy with a reduced coefficient of thermal expansion by an order of magnitude. These characteristics make them suitable for use in a myriad of applications, including medical, electronics, aerospace engineering, telecommunications and cryogenic components.

Iron-nickel alloys demonstrate a very low expansion rate at room temperature. These low thermal expansion alloys have been used in modern applications which require joining of a metal to glass or ceramics and in areas that demand same thermal expansion rates for materials to be joined in order to prevent associated problems within the joint area.

Alloy K, also known as Nilo K and Kovar, consists of 29% nickel and 17% cobalt. Its composition is strictly controlled within narrow limits to ascertain accurate and universal thermal expansion and mechanical properties. This facilitates the alloy provide high integrity ceramic-to-metal and glass-to- metal seals in applications demanding thermal shock resistance and reliability.

Conditioning the alloy utilizing thermal degassing / decarburising in a moist hydrogen atmosphere is recommended so as to attain optimum conditions for subsequent sealing operations. The conditioned material is then subjected to a controlled oxidation heat treatment to form a tightly adherent surface oxide film of optimum thickness and nature.

Properties of Alloy K

The following are the typical physical properties of Alloy K:

Coefficient of thermal expansion	4.6-5.2 at 30-400 °C; 5.1-5.5 at 30-450 °C
Density	8.25 g/cm^3
Melting point	1450 °C
Curie point	425 °C
Thermal conductivity	17.0 W/m°C
Specific heat	0.50 J/g.°C
Young's modulus (MPa)	139,000

Applications of Alloy K

Alloy K is often utilized with both alumina type ceramic materials and the harder Pyrex or borosilicate glasses in the production of hermetic seals. It is widely used in the manufacture of diodes and transistors for lids and closures in many different hybrid electronic circuit packages as well as in the production of microwave tubes.

Pig Iron

Pig iron is the product of smelting iron ore (also ilmenite) with a high-carbon fuel and reductant such as coke, usually with limestone as a flux. Charcoal and anthracite are also used as fuel and reductant.

Pig iron is produced by smelting or iron ore in blast furnaces or by smelting ilmenite in electric furnaces.

The vast majority of pig iron is produced and consumed within integrated steel mill complexes. In this context the term "pig iron" is something of a misnomer: within integrated steel mills, blast furnace iron is transferred directly to the steel plant in liquid form, better known as "hot metal" or "blast furnace iron".

The term "pig iron" dates back to the time when hot metal was cast into ingots before being charged to the steel plant. The moulds were laid out in sand beds such that they could be fed from a common runner. The group of moulds resembled a litter of sucking pigs, the ingots being called "pigs" and the runner the "sow".

Merchant Pig Iron

Merchant pig iron is cold pig iron, cast into ingots and sold to third parties as feedstock for the steel and ferrous casting industries.

Merchant pig iron is produced by:

- Dedicated merchant plants: All of whose production is sold to external customers.
- Integrated steel mills: With iron that is surplus to their internal requirements and cast into ingots and sold to the merchant market.

Types of Merchant Pig Iron

Merchant pig iron comprises three main types:

- Basic pig iron: Used mainly in electric arc steelmaking.
- Foundry pig iron (also known as haematite pig iron): Used in mainly in the manufacture of grey iron castings in cupola furnaces.
- High purity pig iron (also known as nodular pig iron): Used in the manufacture of ductile [also known as nodular or spheroidal graphite – SG] iron castings.

There are also various sub-types, for example low manganese basic pig iron, semi-nodular pig iron etc.

Composition and Characteristics

Pig iron contains at least 92% Fe and has a very high carbon content, typically 3.5 - 4.5%.

Other constituents are given in the table below:

Typical pig iron characteristics (% by weight)					
Pig iron type	C	Si	Mn	S	P
Basic	3.5 - 4.5	≤1.25	≤1.0	≤0.05	0.08-0.15
Foundry	3.5 - 4.1	2.5 - 3.5	0.5 - 1.2	≤0.04	≤0.12
High Purity/ Nodular	3.7 - 4.7	0.05 -1.5	≤0.05	≤0.025	≤0.035

Pig iron is supplied in a variety of ingot sizes and weights, ranging from 3 kg up to more than 50 kg.

Pig Iron Production

Blast furnace economics are such that larger units have lower unit production costs, hence there has been a trend to bigger and bigger furnaces. Modern blast furnaces produce more than 10,000 tonnes per day.

Modern blast furnaces produce more than 10,000 tonnes per day.

Blast Furnace Process

The blast furnace is a counter-current gas/solids reactor in which the descending column of burden materials [coke, iron ore and fluxes/additives] reacts with the ascending hot gases. The process is continuous with raw materials being regularly charged to the top of the furnace and molten iron and slag being tapped from the bottom of the furnace at regular intervals.

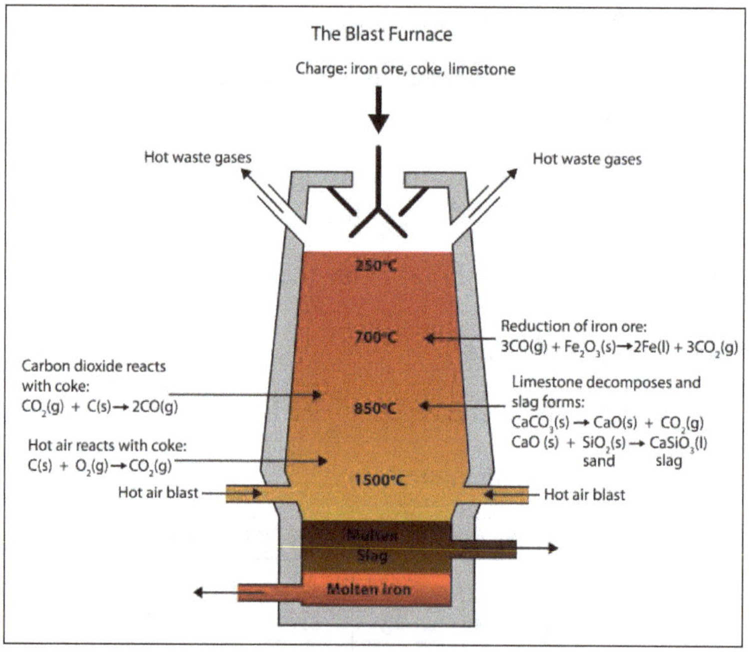

Key steps of the process are as follows:

- Upper part of the furnace - Free moisture is driven off from the burden materials and hydrates and carbonates are disassociated.

- Lower part of the blast furnace shaft - Indirect reduction of the iron oxides by carbon monoxide and hydrogen occurs at 700-1,000 °C.
- Bosh area of the furnace where the burden starts to soften and melt - Direct reduction of the iron [and other] oxides and carbonization by the coke occurs at 1,000-1,600 °C. Molten iron and slag start to drip through to the bottom of the furnace [the hearth].

Between the bosh and the hearth are the tuyeres [water cooled copper nozzles] through which the blast - combustion air, preheated to 900-1,300 °C, often enriched with oxygen - is blown into the furnace. Immediately in front of the tuyeres is the combustion zone, the hottest part of the furnace, 1,850-2,200 °C, where coke reacts with the oxygen and steam in the blast to form carbon monoxide and hydrogen [as well as heat] and the iron and slag melt completely.

Molten iron and slag collect in the furnace hearth. Being less dense, the slag floats on top of the iron. Slag and iron are tapped at regular intervals through separate tap holes. For merchant pig iron production, the iron is cast into ingots; in integrated steel mills, the molten iron or hot metal is transferred in torpedo ladle cars to the steel converters. Slag is transferred to slag pits for further processing into usable materials, for example raw material for cement production, road construction, etc.

Pig iron casting machine.

Blast Furnace Reactions

The principal reactions are:

$$2C + O_2 \rightarrow 2CO$$

$$C + H_2O \rightarrow CO + H_2$$

$$CO_2 + C \rightarrow 2CO$$

$$3Fe_2O_3 + CO \rightarrow CO_2 + 2Fe_3O_4$$

$$Fe_3O_4 + CO \rightarrow CO_2 + 3FeO$$

$$FeO + CO \rightarrow Fe + CO_2$$

When charging the blast furnace, burden materials are added in layers. Charging is done either with an elevator in which a bucket is lifted up and set down on the top of the furnace o be emptied directly into the furnace [bell system] or by conveyor belts to the top of the furnace where materials are charged into a bin fixed to the top of the furnace [bell-less system] and from there into the furnace. By means of a rotating chute it is possible to achieve very uniform distribution of the charge materials across the furnace. The bell-less system has the additional advantage that less energy rich blast furnace gas is lost during charging.

The additives and fluxes serve to convert the waste or gangue materials in the charge [mainly silica and alumina] into a low melting point slag which also dissolves the coke ash and removes sulphur. For example:

$$CaCO_3 \rightarrow CaO + CO_2$$

$$CaO + SiO_2 \rightarrow CaSiO_3$$

$$FeS + CaO + C \rightarrow CaS + FeO + CO$$

The blast furnace itself is a steel shaft lined with fire resistant, refractory materials. The hottest part of furnace - where the walls reach a temperature >300°C - are water cooled. The whole structure is supported from the outside by a steel frame.

The blast furnace gas that leaves the top of the furnace is a mixture of carbon dioxide, carbon monoxide, hydrogen and nitrogen and has a calorific value between 3,200 and 4,000 kJ/m^3. After cleaning, it is used for a variety of purposes, including heating of the hot blast stoves ["cowpers"], in iron ore agglomeration plants and for electricity generation. The credit for this gas is an important factor in keeping blast furnace operating costs low.

Steel

Steel, the world's foremost construction material, is an iron alloy that contains between 0.2% and 2% carbon by weight.

Production

Today, most steel is produced by basic oxygen methods (also known as basic oxygen steelmaking or BOS). BOS derives its name from the process that requires oxygen to be blown into large vessels containing molten iron and scrap steel.

Although BOS accounts for the largest share of global steel production, the use of electric arc furnaces (EAF) has been growing since the early 20th century and now accounts for about one-third of all steel production.

Grades

According to the World Steel Association, there are over 3,500 different grades of steel, encompassing unique physical, chemical, and environmental properties. You might be wondering how there can be so many different grades of steel when steel is only composed of iron and carbon. That's because, the amount of carbon, as well as the level of impurities and additional alloying elements, determine the properties of each steel grade, and those combinations reach very high numbers. Commercial steel is generally classified into four groups depending on their metal alloy content and end-use applications:

- Carbon Steels (which include low carbon, medium carbon, and high carbon steels).
- Alloy Steels (which include such common alloy metals as manganese, silicon, nickel, and chromium).
- Stainless Steels (which contain about 10 percent chromium and are classified as austenitic, ferritic, and martensitic).
- Tool Steels (which are steels that are alloyed with high temperature and hard metals, such as molybdenum and tungsten).

Steel has a wide range of alloy compositions, which allow it to have diverse mechanical properties.

Properties

Different types of steel are produced according to the properties required for their application, and various grading systems are used to distinguish steels based on these properties, which include density, elasticity, melting point, thermal conductivity, strength, and hardness (among others). To make different steels, manufacturers vary the type and quantity of alloy metals, the production process, and the manner in which the steels are worked to produce particular products.

Steels can be broadly categorized into four groups based on their chemical compositions:

- Carbon Steels.
- Alloy Steels.
- Stainless Steels.
- Tool Steels.

Applications

It is steel's properties that have made it the most widely used and most recycled metal material on earth. From stainless and high-temperature steels to flat carbon products, steel's various forms and alloys offer different properties to meet a wide range of applications.

Carbon Steel

Carbon steel is the most widely used engineering and construction material for industrial applications on a large scale, including marine structures, power plants, transportation, chemical processing and petroleum production and refining.

Low Carbon Steel

Low carbon steels such as 302, 304 or 316 grades of stainless are typically used in applications which require high degrees of corrosion resistance but do not require a hardened surface. The carbon content of these steels typically range between 0.03-0.08%, and consumers typically use these grades of stainless (often without thinking about it) in kitchen equipment, silverware or almost any grade of un-plated steel used in food preparation. It's great because it can survive the dishwasher without rusting, but it cannot be case hardened due to the very low carbon content.

While it can be used to make linear shafting, it isn't suitable for loaded ball contact. So if a linear ball bushing were to be used on a soft 304 stainless steel shaft, for example, the balls in the bearing would quickly impact the shaft surface, resulting in visible ball tracking on its surface and a drastic reduction in both bearing and shaft life. It can, however, be used in conjunction with polymer, plain-style bearings which provide great options for both corrosion resistance and self-lubrication. For the right applications, a 300 series stainless steel linear shaft presents a great option for withstanding tough environmental conditions.

Medium Carbon Steel

Medium carbon steels include grades with carbon contents ranging from 0.25% to 0.60% of the steel mass. Medium carbon grades are typically employed in conjunction with alloys such as chromium, nickel and molybdenum to produce high strength, wear resistance and toughness. Products using medium grades of carbon steel include gears, axles, studs and other machine components that require optimal combinations of strength and toughness.

Medium carbon steels have good machining characteristics, and one of the more popular grades used in machined steel product is AISI 1045. AISI 1045 can also be hardened by heating the material too approximately 820-850 °C (1508 -1562 °F) and held until the material reaches a uniform temperature. It should be soaked for one hour per 25 mm section of material and subsequently cooled in still air.

High Carbon Steel

High carbon steels are those with carbon contents between 0.60% and 1.4% of the overall weight. The alloys in this particular category constitute the strongest and hardest within the three groups, but they are also the least ductile. These steels are used in a range of different mechanical, cutting and bearing applications as it can be hardened through heat treating and tempering. Additional alloys can be added to this steel category in order to generate different characteristics. Chromium and Manganese, for example, are used in the composition of 52100 steel and aid in the hardening process while enhancing the steel's resistance to corrosion. Since 52100 is one of the steel grades frequently used to manufacture linear shafting, precise control of the case depth can be critical to generate a shaft with both a hardened surface (for loaded ball contact) and an un-hardened inner core which prevents the shaft from becoming brittle.

Steel alloys are given designators by organizations such as the American Iron and Steel Institute (AISI) and the American Society for Testing and Materials (ASTM) for easier classification and identification. AISI typically follows a four digit system, where the first two digits indicate the alloy, and the second two digits denote the carbon content. ASTM uses an "A" to denote ferrous materials, followed by an arbitrarily assigned number for each alloy.

Regardless of the system used, this standardization allows cross talk between designers, engineers, and builders to ensure the proper material is being selected and used in engineering projects. It also makes looking up physical properties of alloys very easy, as a simple search with the identification number produces the correct information.

Alloy Steel

Alloy steel is a class of steel that, in addition to carbon, is alloyed with other elements, ranging from 1 wt.% to 50 wt.%, which are used to enhance the material's various properties.

These elements commonly include manganese, nickel, chromium, molybdenum, vanadium, silicon, and boron. Less common elements include aluminium, cobalt, copper, cerium, niobium, titanium, tungsten, tin, zinc, lead, and zirconium.

Types of Alloy Steel

There are multiple subcategories of alloy steel. These include:

- Low-alloy steel.
- High-strength low alloy (HSLA) steel.
- High-alloy steel.
- Stainless steel.
- Microalloyed steel.
- Advanced high-strength steel (AHSS).
- Maraging steel.
- Tool steel.

Low alloy steels generally contain less than 8 wt.% non-iron elements, whereas high-alloy steels contain more than 8 wt.% non-iron elements. Both typically have superior mechanical properties in comparison to carbon steels.

Properties of Alloy Steel

Alloy steels can contain a wide variety of elements, each of which can enhance various properties of the material, such as mechanical thermal and corrosion resistance. Elements added in low quantities of less than around 5 wt.% tend to improve mechanical properties, for example increasing hardenability and strength, whereas larger additions of up to 20 wt.% increase corrosion resistance and stability at high or low temperatures.

The effects of adding various elements to steel, along with the typical amounts in weight fraction, is summarised in the table below.

Element	Symbol	wt. %	Function
Aluminium	Al	0.95–1.30	Alloying element in nitriding steels.
Bismuth	Bi	–	Improves machinability.
Boron	B	0.001–0.003	Improves hardenability.
Chromium	Cr	0.5–2.0	Improves hardenability.
		4–18	Corrosion resistance.
Copper	Cu	0.1–0.4	Corrosion resistance.
Lead	Pb	–	Improves machinability.
Manganese	Mn	0.25–0.40	Prevents brittleness in combination with sulfur.
		>1	Increases hardenability.
Molybdenum	Mo	0.2–0.5	Inhibits grain growth.
Nickel	Ni	2–5 12–20	Increases toughness. Improves corrosion resistance.
Silicon	Si	0.2–0.7	Increases strength and hardenability.
		2	Increases yield strength (spring steel).
		Higher %	Increases magnetic properties.
Sulfur	S	0.08–0.15	Improves machinability (free-machining steel properties).
Titanium	Ti	–	Reduces martensitic hardness in Cr steels.
Tungsten	W	–	Increases hardness at high temperatures.
Vanadium	V	0.15	Increases strength while maintaining ductility, promotes fine grain structure.

Overall, in comparison to carbon steels, alloy steels can exhibit increased strength, ductility and toughness. The disadvantages, however, are that alloy steels usually have lower machinability, weldability and formability.

Production and Processing

The alloying and processing methods for alloy steel depend on the desired result. The required combination of elements is first melted together in a furnace at over 1600 °C for 8 to 12 hours. The steel is then annealed at over 500 °C in order to remove impurities and to alter the physical and chemical properties.

Next, mill scale (a mixture of iron oxides), which results from the annealing process, is removed from the surface of the steel with hydrofluoric acid before repeating the annealing and descaling process. Finally, the steel is melted and cast into for rolling and shaping into the final form.

Applications and Examples

As the term alloy steel encompasses numerous types of steel, its application area is broad.

Low alloy steels are used in a wide range of industries due to their extreme strength, machinability, cost-effectiveness and availability. They are found in military vehicles, construction equipment, ships, pipelines, pressure vessels oil drilling platforms and in structural components. Examples include HY80 and HY100.

High-alloy steels can be expensive to manufacture and difficult to process. Nevertheless, their superior hardness, toughness and corrosion resistance make them ideal for structural components, automotive applications, chemical processing and power generating equipment. Examples of high-alloy steels include the grades HE, HF, HH, HI, HK, and HL.

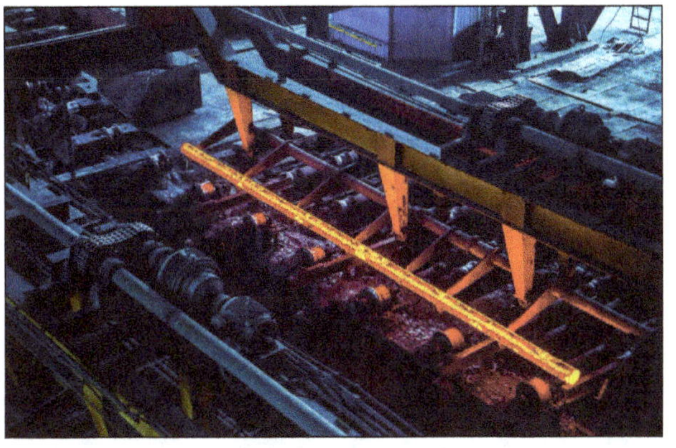

Stainless Steel

Stainless steel is a class of iron-based alloys with a minimum chromium content of 10.5 wt.%. It is characterised by its superior resistance to corrosion in comparison to other steels.

Properties of Stainless Steel

Corrosion Resistance

The corrosion resistance of stainless steel is mostly a result of the chromium content. A stable layer of chromium oxide is formed on the surface of the steel, which prevents chemical reactions with the bulk of the material. This oxide layer is extremely thin, 2 nm – 3 nm in thickness, is passive (highly resistant to corrosion), tenacious (adheres well to the bulk) and self-healing (reforms when cracked or damaged).

Nevertheless, stainless steel can corrode under certain circumstances. Uniform corrosion can occur when exposed to acidic solutions such as strong sulfuric acid and hydrochloric acid, and to basic solutions such as sodium hydroxide. Localised corrosion can occur in the form of pitting corrosion, crevice corrosion and stress corrosion cracking, for example when exposed to chloride ions.

Mechanical Properties

As the term stainless steel encompasses a wide range of materials, the mechanical properties, of course, are quite diverse. In general, the values tested for include yield strength, tensile strength, ductility, hardness, toughness, creep resistance and fatigue resistance. Specific values can be found on Matmatch for thousands of different stainless steels.

Electric and Magnetic Properties

Stainless steel is a relatively poor conductor of electricity. The electrical conductivity of 18 wt.% Cr. 8 wt.% Ni stainless steel at 20 °C is 1.45 × 106 S/m compared to 5.96 × 107 S/m for pure copper.

Ferritic, martensitic and duplex stainless steels are classified as magnetic, whereas austenitic is classified as non-magnetic. An important value often used to quantify this is the relative magnetic permeability μr, which is related to the magnetic susceptibility χm via the equation m=r-1. Magnetic stainless steels typically have permeability values of around 14, whereas non-magnetic stainless steels are close to the minimum value of 1.

Recycling

All stainless steels are completely recyclable. As they contain significant amounts of valuable elements such as chromium and nickel, recycling of scrap stainless steel is in fact extremely cost-effective. Modern stainless steels are usually manufactured with 60% scrap, including reclaimed scrap, such as from consumer products and industrial equipment, and industrial scrap, such as offcuts from the manufacturing process.

Categories of Stainless Steel

Stainless steels are classified into four main categories according to their crystal structure: ferritic, austenitic, martensitic and duplex.

Ferritic

Ferritic stainless steels possess a body-centred cubic crystal structure, similar to that of pure iron. They generally have a high chromium content of between 10.5 wt.% and 18 wt.%. They have a low carbon content, which gives them a relatively low strength and contain little to no nickel. Ferritic stainless steels are not hardenable by heat treatment. After annealing, they have yield strengths in the range of 275 MPa – 350 MPa.

Ferritic stainless steels are known to have low ductility, resulting in reduced formability, are magnetic, have limited toughness and often lower corrosion resistance compared to other stainless steels. Their absence of nickel, however, gives them a lower cost.

One of the main advantages of ferritic stainless steels is their high resistance to stress corrosion cracking.

Examples and Applications of Ferritic Stainless Steels

- 409, used in the automotive industry for parts such as exhaust systems and tailpipes.
- 410S, used in the oil and gas industry, ore processing and thermal processing.
- 430, used in dishwasher linings, refrigerator cabinet panels, element supports and fasteners, stove trim rings and chimney liners and automotive trim and lashing wires.

Austenitic

Austenitic stainless steels are the most common category of stainless steel. They possess a face-centred cubic crystal structure. In addition to the main alloying elements of iron and chromium, these steels also contain nickel, manganese and nitrogen.

They are not hardenable via heat treatment but only by cold working (also known as work hardening). These steels have relatively low strength compared to the other steels and lower resistance to corrosion.

Austenitic stainless steels are split into two subcategories: The 300 series and 200 series. For the 300 series, the austenitic structure is achieved via the addition of nickel, whereas for the 200 series, this is achieved primarily via the addition of manganese and nitrogen.

Austenitic stainless steels are non-magnetic, however, the 300 series can become magnetic after cold working. The nickel content in the 300 series makes it suitable for low-temperature cryogenic applications.

Examples and Applications of Austenitic Stainless Steels

- 304, the most common stainless steel contains 18 wt.% chromium and 8 wt.% nickel and therefore is often referred to as 18/8. It is used for many applications including cutlery, kitchen equipment, food processing equipment, automotive and aerospace structural components and marine fasteners.
- 316, the second most common stainless steel, is used for example in food preparation equipment, chemical processing equipment, laboratory benches and equipment, boat fittings, heat exchangers and pharmaceutical and textile equipment and surgical equipment.
- 317, a low-carbon stainless steel with increased chromium, nickel and molybdenum for greater corrosion resistance. This is used for paper handling equipment, chemical and petrochemical processing equipment, condensers in power generation stations, food processing equipment and textile equipment.

Martensitic

Martensitic stainless steels posses the body-centred tetragonal crystal system. They can range from having a low to high carbon content, up to 1.2 wt.%, contain 12 wt.% – 15 wt.% chromium and 0.2 wt.% to 1.0 wt.% molybdenum. They do not contain nickel. Due to the presence of carbon, they are hardenable via heat treatment, similarly to carbon steels.

Martensitic stainless steels are magnetic and possess relatively high ductility and toughness, making them easier to form. They can be moderately hardened via cold working. After annealing, they typically have yield strengths of around 275 MPa.

This is dependant on carbon content: higher carbon content results in increased strength and hardness but decreased ductility and toughness. They exhibit moderate corrosion resistance and poor weldability.

Examples and Applications of Martensitic Stainless Steels

- 403, used in compressor blades and turbine parts.

- 410, used for highly stressed parts, for examples in turbine blades, kitchen utensils, bolts, nuts and screws, pump and valve parts, dental and surgical instruments, nozzles and parts for oil well pumps.
- 416, which has the highest machinability of any stainless steel, is used for electrical motors, nuts and bolts, pumps, valves, automatic screw machine parts and gears.
- 420, used for cutlery, knife blades, surgical instruments, needle valves, shear blades, scissors and hand tools.

Duplex

Duplex stainless steels possess a mixed microstructure of ferrite and austenite. They contain a high chromium content of 22 wt.% – 25 wt.%, a molybdenum content of up to 5 wt.% and a low nickel content.

Duplex stainless steels exhibit roughly twice the strength of austenitic stainless steels and possess greater resistance to chloride stress corrosion cracking compared to austenitic stainless steels, albeit less so than ferritic. They exhibit toughness between that of ferritic and austenitic.

Examples and Applications of Duplex Stainless Steels

Due to these properties, duplex stainless steels are ideal for demanding environments where good mechanical properties are required in addition to good corrosion resistance. Common examples of duplex stainless steels include:

- 2205, used in chemical processing, oil and gas processing equipment, marine and other high chloride environments, effluent scrubbing systems, pulp and paper industry, cargo tanks for ships and trucks, food processing equipment, biofuels plant.
- 2304, used in chloride containing environments, welded pipe systems, transportation, heat exchanger tubes, construction, pressure vessels, caustic solutions and the food industry.
- 2507, used in oil and gas industry equipment, offshore platforms, chemical process industries, desalination plants, mechanical and structural components and power industry FGD systems.

Tool Steel

Tool steel refers to a variety of carbon and alloy steels that are particularly well-suited to be made into tools. Their suitability comes from their distinctive hardness, resistance to abrasion and deformation and their ability to hold a cutting edge at elevated temperatures. As a result tool steels are suited for their use in the shaping of other materials.

There are six groups of tool steel: water-hardening, cold-work tool steels, shock-resisting, high-speed, hot-work, and special purpose/plastic mold tools steel. The choice of group to select depends on, cost, working temperature, required surface hardness, strength, shock resistance, and toughness requirements. The more severe the service condition (higher temperature, abrasiveness, corrosiveness, loading), the higher the alloy content and consequent amount of carbides required for the tool steel.

Water-hardening Group

Named from its essential property of having to be water quenched. This group of tool steel is essentially plain high carbon steel. It is commonly used because of its low cost.

W-group tool steel gets its name from its defining property of having to be water quenched. W-grade steel is essentially high carbon plain-carbon steel. This group of tool steel is the most commonly used tool steel because of its low cost compared to others. They work well for small parts and applications where high temperatures are not encountered; above 150 °C (302 °F) it begins to soften to a noticeable degree. Its hardenability is low, so W-group tool steels must be subjected to a rapid quenching, requiring the use of water. These tool steels can attain high hardness (above HRC 66) and are rather brittle compared to other tool steels. W-steels are still sold, especially for springs, but are much less widely used than they were in the 19th and early 20th centuries. This is partly because W-steels warp and crack much more during quench than oil-quenched or air hardening steels.

Cold-work Group

This is a group of three tool steels: oil-hardening, air-hardening, and high carbon-chromium. The steels in the group have high hardenability and wear resistance, with average toughness. Typically they are in the production of larger parts or parts that have a minimum distortion requirement when being hardened.

Both Oil quenching and Air-hardening both reduce the distortion and higher stress caused by the quick water quenching. Because of this they are less likely to crack.

- Oil-hardening

  A very common oil hardening steel is O1 steel. It is a very good cold work steel and also makes very good knives and forks. It can be hardened to about 57-61 HRC.

- Air-hardening

  The first air-hardening grade tool steel was mushet steel, which was known as air-hardening steel at the time.

  Modern air-hardening steels are characterized by low distortion during heat treatment because of their high-chromium content. Their machinability is good and they have a balance of wear resistance and toughness (i.e. between the D- and shock-resistant grades).

- High carbon-chromium.

D-type

The D-type, of the cold-work class of tool steels, contain between 10% and 13% chromium. These steels retain their hardness up to a temperature of 425 °C (797 °F). Common applications for these tool steels include forging dies, die-casting die blocks, and drawing dies. Due to their high chromium content, certain D-type tool steels are often considered stainless or semi-stainless, however their corrosion resistance is very limited due to the precipitation of the majority of their chromium and carbon constituents as carbides.

D2 Tool Steel is very wear resistant but not as tough as lower alloyed steels. The mechanical properties of D2 are very sensitive to heat treatment. It is widely used for the production of shear blades, planer blades and industrial cutting tools; sometimes used for knife blades.

A	A2 – A10	Air hardening, Medium alloys
D	D2 – D7	High carbon, high chromium
O	O1 – O7	Oil hardening, Low carbon

Type

ISO 1.2767, also known as DIN X 45 NiCrMo 4, AISI 6F7, and BS EN 20 B, is an air-hardening tool steel with a primary alloying element of nickel. It possesses good toughness, stable grains, and is highly polishable. It is primarily used for dies in plastic injection molding application that involve high stresses. Other applications include blanking dies, forging dies, and industrial blades.

The following steel grades are also in cold work steel group.

AISI Code	AISI Designation	Type of Tool steel
W	W1 to W7	High carbon water hardening steel
	W1 A – 1B	Carbon

	W2 – W3	Carbon Vanadium
	W4 – W5	Carbon Chromium
	W7	Carbon Chromium Vanadium

Shock-resisting Group

This class has high shock resistance and good hardenability. It is designed to resist shock at both low and high temperatures. It also has a very high impact toughness and relatively low abrasion resistance.

Carbide-forming alloys provide the necessary abrasion resistance, hardenability, and hot-work characteristics. This family of steels displays very high impact toughness and relatively low abrasion resistance and can attain relatively high hardness (HRC 58/60). In the US, toughness usually derives from 1 to 2% silicon and 0.5-1% molybdenum content. In Europe, shock steels often contain 0.5-0.6 % carbon and around 3% nickel. 1.75% to 2.75% nickel is still used in some shock resisting and high strength low alloy steels (HSLA), such as L6, 4340, and Swedish saw steel, but it is relatively expensive. An example of its use is in the production of jackhammer bits.

High Speed Group

T-type and M-type tool steels are used for cutting tools when strength and hardness must be retained at high temperatures.

High-speed steel (HSS or HS) is a subset of tool steels, commonly used in tool bits and cutting tools. It is often used in power-saw blades and drill bits. It is superior to the older high-carbon steel tools used extensively through the 1940s in that it can withstand higher temperatures without losing its temper (hardness). This property allows HSS to cut faster than high carbon steel, hence the name high-speed steel. At room temperature, in their generally recommended heat treatment, HSS grades generally display high hardness (above HRC60) and abrasion resistance (generally linked to tungsten and vanadium content often used in HSS) compared with common carbon and tool steels.

M	M1, M7, M10	Molybdenum
	M30, M33, M34, M42, M43, M46, M47	Molybdenum, Cobalt
	M2, M3, M4	Molybdenum, Tungsten
	M6, M15, M35, M36< M41, M44, M45	Molybdenum, Tungsten, Cobalt
T	T1, T2, T3, T7, T9	Tungsten
	T4, T5, T6, T8, T15	Tungsten, cobalt

Hot-working Group

H-group tool steels were specifically developed to maintain strength and hardness while exposed to prolonged elevated temperatures.

Hot-working steels are a group of steel used to cut or shape material at high temperatures. H-group tool steels were developed for strength and hardness during prolonged exposure to elevated temperatures. These tool steels are low carbon and moderate to high alloy that provide good hot hardness and toughness and fair wear resistance due to a substantial amount of carbide. H1 to H19 are based on a chromium content of 5%; H20 to H39 are based on a tungsten content of 9-18% and a chromium content of 3–4%; H40 to H59 are molybdenum based.

H	H 10, H11, h12, H13	Chromium, Molybdenum
	H14, H16, H19, H23	Chromium, Tungsten
	H20, H21, H22, H24, H25, H26	Tungsten
	H15, H41, H42, H43	Molybdenum

Special Purpose Group/Plastic Mold Steel

P-type tool steel is short for plastic mold steels. They are designed to meet the requirements of zinc die casting and plastic injection molding dies. Common steel grades like P20, 420 etc.

L-type tool steel is short for low alloy special purpose tool steel. L6 is extremely tough.

F-type tool steel is water hardened and substantially more wear resistant than W-type tool steel.

F	F1	High carbon, low alloys
	F2, F3	Tungsten
L	L1, L3, L7	Carbon > 0.65%, Chromium
	L2	Carbon <0.65%, Chromium
	L6	Carbon > 0.65%, Nickel
S	S1, S3	Tungsten
	S2, S4, S5, S6	Silicon
	S7	Chromium
P	P1- P20, P21	Low carbon mold steel

Tool steels are metallurgically “clean,” high-alloy steels that are melted in relatively small heats in electric furnaces and produced with careful attention to homogeneity. They can be further refined by argon/oxygen decarburization (AOD), vacuum methods, or electroslag refining (ESR). As a result, tool steels are often specified for critical high-strength or wear-resistant applications. Because of their high alloy content, tool steels must be rolled or forged with care to produce satisfactory bar products.

To develop their best properties, tool steels are always heat treated. Because the parts may distort during heat treatment, precision parts should be semifinished, heat treated, then finished. Severe distortion is most likely to occur during liquid quenching, so an alloy should be selected that provides the needed mechanical properties with the least severe quench.

References

- Iron: newworldencyclopedia.org, Retrieved 26 March, 2019
- Iron-alloy: themetalcasting.com, Retrieved 02 May, 2019
- Cast-iron-alloy: themetalcasting.com, Retrieved 18 April, 2019

- Pig-iron: metallics.org, Retrieved 03 June, 2019
- Metal-profile-steel-2340175: thebalance.com, Retrieved 15 July, 2019
- The-difference-between-carbon-steels: blog.misumiusa.com, Retrieved 12 February, 2019
- Alloy-steel: matmatch.com, Retrieved 16 July, 2019
- What-is-tool-steel, steel-knowledge: astmsteel.com, Retrieved 30 March, 2019

3

Aluminium Alloys: Properties and Applications

Aluminium is a silvery-white, soft, non-magnetic and ductile metal which belongs to the boron group. Its alloys are known for their improved strength, electrical conductivity and resistance to corrosion. All the diverse properties and applications of aluminium alloys have been carefully analyzed in this chapter.

Aluminium

Aluminum (or aluminium) (chemical symbol Al, atomic number is 13) is a soft, lightweight metal with a silvery appearance and the ability to resist corrosion. It is the most abundant metallic element in the Earth's crust (estimated at between 7.5 and 8.1 percent). The free element, rarely found in nature, occurs in oxygen-deficient environments such as volcanic mud. Its main ore is bauxite. Whether measured in terms of quantity or value, the global use of aluminum exceeds that of any other metal except iron, and it is important in virtually all segments of the world economy.

Structural components made from aluminum and its alloys are vital to the aerospace industry and very important in other areas of transportation and building. In addition, aluminum and its alloys are used in packaging, cooking utensils, electrical transmission lines, water purification processes, electronic devices and compact discs, paint, and pyrotechnics. Aluminum compounds also serve a wide variety of purposes. For instance, aluminum ammonium sulfate is a mordant for dyeing, and is used in water purification and sewage treatment; aluminum acetate solution is an astringent; aluminum chloride is used in paints and anti-perspirants; and aluminum borate, phosphate, and fluorosilicate are used in the production of glass and ceramics. Yet, aluminum is one of the few abundant elements that appear to have no beneficial biological role; a small percentage of people are allergic to it.

Notable Characteristics

Physical Properties

In the periodic table, aluminum is located in group 13 (former group 3A), between boron and gallium. In addition, it lies in period 3, between magnesium and silicon. It is considered a member of the "poor metal" group of chemical elements. It is nontoxic, nonmagnetic, and nonsparking. The atoms in the metal are arranged in a face-centered cubic structure.

Physical properties	
Phase	solid
Density (near r.t.)	2.70 $g \cdot cm^{-3}$

Liquid density at m.p.	2.375 $g \cdot cm^{-3}$
Melting point	933.47 K (660.32 °C, 1220.58 °F)
Boiling point	2792 K (2519 °C, 4566 °F)
Heat of fusion	10.71 $kJ \cdot mol^{-1}$
Heat of vaporization	294.0 $kJ \cdot mol^{-1}$
Heat capacity	(25 °C) 24.200 $J \cdot mol^{-1} \cdot K^{-1}$

Vapor pressure						
P/Pa	1	10	100	1 k	10 k	100 k
at *T*/K	1482	1632	1817	2054	2364	2790

Aluminum is one of the few metals that retain full silvery reflectance in finely powdered form, making it an important component of silver paints. Pure aluminum serves as an excellent reflector (approximately 99%) of visible light and a good reflector (approximately 95%) of infrared. It is a good thermal and electrical conductor, by weight better than copper. It is capable of being a superconductor, with a superconducting critical temperature of 1.2 Kelvin.

This metal has about one-third the density and stiffness of steel. It is ductile, and easily machined, cast, and extruded. The yield strength of pure aluminum is 7-11 MPa, while aluminum alloys have yield strengths ranging from 200 to 600 MPa. Also, pure aluminum has a low tensile strength, but its alloys display a marked improvement in mechanical properties, especially when tempered.

The statue Eros in Piccadilly Circus,
and is one of the first statues to be cast in aluminum.

Atomic properties	
Crystal structure	face centered cubic 0.4032 nm
Oxidation states	3 (amphoteric oxide)

Electronegativity	1.61 (Pauling scale)
Ionization energies (more)	1st: 577.5 kJ·mol^{-1}
	2nd: 1816.7 kJ·mol^{-1}
	3rd: 2744.8 kJ·mol^{-1}
Atomic radius	125 pm
Atomic radius (calc.)	118 pm
Covalent radius	118 pm

Chemical Properties

Aluminum is highly resistant to corrosion, due to a thin surface layer of aluminum oxide that forms when the metal is exposed to air, effectively preventing further oxidation. The strongest aluminum alloys are less corrosion resistant due to galvanic reactions with alloyed copper.

When combining with other elements, aluminum can have different oxidation states: +1, +2, and +3. Of these, the +3 oxidation state is most common.

Oxidation state one:

- AlH is produced when aluminum is heated at 1500 °C in an atmosphere of hydrogen.
- Al_2O is made by heating the normal oxide, Al_2O_3, with silicon at 1800 °C in a vacuum.
- Al_2S can be made by heating Al_2S_3 with aluminum shavings at 1300 °C in a vacuum. It quickly breaks up to regenerate the starting materials. The selenide is made in a parallel manner.
- AlF, AlCl, and AlBr exist in the gaseous phase when the corresponding tri-halide is heated with aluminum.

Oxidation state two:

- Aluminum monoxide, AlO, is present when aluminum powder burns in oxygen.

Oxidation state three:

- According to Fajans' rules, the simple trivalent cation Al^{3+} is not expected to be found in anhydrous salts or binary compounds such as Al_2O_3. The hydroxide is a weak base and aluminum salts of weak acids, such as carbonate, can't be prepared. The salts of strong acids, such as nitrate, are stable and soluble in water, forming hydrates with at least six molecules of water of crystallization.
- Aluminum hydride, $(AlH_3)_n$, can be produced from trimethylaluminum and an excess of hydrogen. It burns explosively in air. It can also be prepared by the action of aluminum chloride on lithium hydride in ether solution, but cannot be isolated free from the solvent.
- Aluminum carbide, Al_4C_3 is made by heating a mixture of the elements above 1000 °C. The pale yellow crystals have a complex lattice structure, and react with water or dilute acids to give methane. The acetylide, $Al_2(C_2)_3$, is made by passing acetylene over heated aluminum.
- Aluminum nitride, AlN, can be made from the elements at 800 °C. It is hydrolyzed by water

to form ammonia and aluminum hydroxide.

- Aluminum phosphide, AlP, is made similarly, and hydrolyses to give phosphine.
- Aluminum oxide, Al_2O_3, occurs naturally as corundum, and can be made by burning aluminum in oxygen or by heating the hydroxide, nitrate or sulfate. As a gemstone, its hardness is only exceeded by diamond, boron nitride, and carborundum. It is almost insoluble in water.
- Aluminum hydroxide may be prepared as a gelatinous precipitate by adding ammonia to an aqueous solution of an aluminum salt. It is amphoteric, being both a very weak acid and forming aluminates with alkalis. It exists in various crystalline forms.
- Aluminum sulfide, Al_2S_3, may be prepared by passing hydrogen sulfide over aluminum powder. It is polymorphic.
- Aluminum iodide, $(AlI_3)_2$, is a dimer with applications in organic synthesis.
- Aluminum fluoride, AlF_3, is made by treating the hydroxide with HF, or can be made from the elements. It consists of a giant molecule which sublimes without melting at 1291 °C. It is very inert. The other trihalides are dimeric, having a bridge-like structure.
- Aluminum fluoride/water complexes: When aluminum and fluoride are together in aqueous solution, they readily form complex ions such as $AlF(H_2O)_5^{+2}$, $AlF_3(H_2O)_3^{0}$, AlF_6^{-3}. Of these, AlF_6^{-3} is the most stable. This is explained by the fact that aluminum and fluoride, which are both very compact ions, fit together just right to form the octahedral aluminum hexafluoride complex. When aluminum and fluoride are together in water in a 1:6 molar ratio, AlF_6^{-3} is the most common form, even in rather low concentrations.
- Organo-metallic compounds of empirical formula AlR_3 exist and, if not also giant molecules, are at least dimers or trimers. They have some uses in organic synthesis, for instance trimethylaluminum.
- Alumino-hydrides of the most electropositive elements are known, the most useful being lithium aluminum hydride, $Li[AlH_4]$. It decomposes into lithium hydride, aluminum and hydrogen when heated, and is hydrolysed by water. It has many uses in organic chemistry, particularly as a reducing agent. The aluminohalides have a similar structure.

This heat sink is made from anodized aluminum.

Clusters

It was reported that clusters of 13 aluminum atoms (Al_{13}) had been made to behave like an iodine atom; and, 14 aluminum atoms (Al_{14}) behaved like an alkaline earth atom. The researchers also bound 12 iodine atoms to an Al_{13} cluster to form a new class of polyiodide. This discovery is reported to give rise to the possibility of a new characterization of the periodic table: Superatoms. The research teams were led by Shiv N. Khanna (Virginia Commonwealth University) and A. Welford Castleman, Jr. (Penn State University).

Isotopes

Aluminum has many isotopes, of which only ^{27}Al (stable isotope) and ^{26}Al (radioactive isotope, $t_{1/2} = 7.2 \times 10^5$ y) occur naturally. The ^{27}Al isotope has a natural abundance of 99.9+ percent. ^{26}Al is produced from argon in the atmosphere by spallation caused by cosmic-ray protons. Aluminum isotopes have found practical application in dating marine sediments, manganese nodules, glacial ice, quartz in rock exposures, and meteorites. The ratio of ^{26}Al to ^{10}Be has been used to study the role of transport, deposition, sediment storage, burial times, and erosion on 10^5 to 10^6 year time scales.

Cosmogenic ^{26}Al was first applied in studies of the Moon and meteorites. Meteorite fragments, after departure from their parent bodies, are exposed to intense cosmic-ray bombardment during their travel through space, causing substantial ^{26}Al production. After falling to Earth, atmospheric shielding protects the meteorite fragments from further ^{26}Al production, and its decay can then be used to determine the meteorite's terrestrial age. Meteorite research has also shown that ^{26}Al was relatively abundant at the time of formation of our planetary system. Many researchers studying meteorites believe that the energy released by the decay of ^{26}Al was responsible for the melting and differentiation of some asteroids after their formation 4.55 billion years ago.

Aluminum Metal Production and Refinement

Aluminum is a reactive metal that is difficult to extract from ore, aluminum oxide (Al_2O_3). Direct reduction—with carbon, for example—is not economically viable since aluminum oxide has a melting point of about 2,000 °C. Therefore, it is extracted by electrolysis; that is, the aluminum oxide is dissolved in molten cryolite and then reduced to the pure metal. By this process, the operational temperature of the reduction cells is around 950 to 980 °C. Cryolite is found as a mineral in Greenland, but in industrial use it has been replaced by a synthetic substance. Cryolite is a mixture of aluminum, sodium, and calcium fluorides: (Na_3AlF_6). The aluminum oxide (a white powder) is obtained by refining bauxite in the Bayer process. (Previously, the Deville process was the predominant refining technology).

The electrolytic process replaced the Wöhler process, which involved the reduction of anhydrous aluminum chloride with potassium. Both of the electrodes used in the electrolysis of aluminum oxide are carbon. Once the ore is in the molten state, its ions are free to move around. The reaction at the cathode (the negative terminal) produces aluminum metal:

$$Al^{3+} + 3\ e^- \rightarrow Al$$

Here, the aluminum ion is reduced (electrons are added). The aluminum metal then sinks to the bottom and is tapped off.

At the positive electrode (anode), oxygen is formed:

$$2\,O^{2-} \rightarrow O_2 + 4\,e^-$$

This carbon anode is then oxidized by the oxygen, releasing carbon dioxide. The anodes in a reduction cell must therefore be replaced regularly, since they are consumed in the process:

$$O_2 + C \rightarrow CO_2$$

Unlike the anodes, the cathodes are not oxidized because there is no oxygen present at the cathode. The carbon cathode is protected by the liquid aluminum inside the cells. Nevertheless, cathodes do erode, mainly due to electrochemical processes. After five to ten years, depending on the current used in the electrolysis, a cell has to be rebuilt because of cathode wear.

Aluminum electrolysis with the Hall-Héroult process consumes a lot of energy, but alternative processes were always found to be less viable economically and/or ecologically. The world-wide average specific energy consumption is approximately 15±0.5 kilowatt-hours per kilogram of aluminum produced from alumina. (52 to 56 MJ/kg). The most modern smelters reach approximately 12.8 kW·h/kg (46.1 MJ/kg). Reduction line current for older technologies are typically 100 to 200 kA. State-of-the-art smelters operate with about 350 kA. Trials have been reported with 500 kA cells.

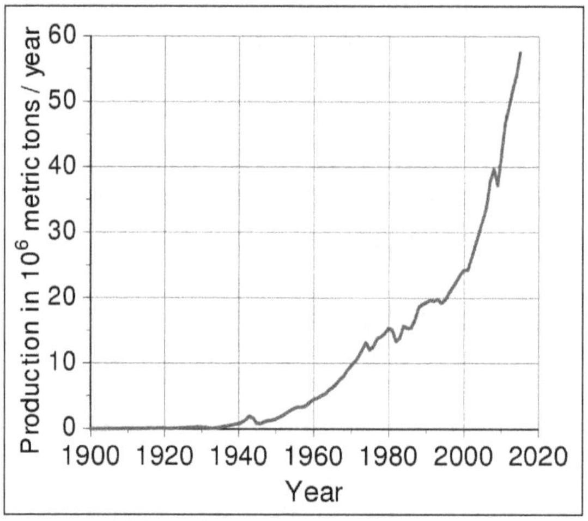

World production trend of aluminum.

Recovery of the metal via recycling has become an important facet of the aluminum industry. Recycling involves melting the scrap, a process that uses only five percent of the energy needed to produce aluminum from ore. However, a significant part (up to 15% of input material) is lost as dross (ash-like oxide). Recycling was a low-profile activity until the late 1960s, when the growing use of aluminum beverage cans brought it to the public consciousness.

Electric power represents about 20 to 40 percent of the cost of producing aluminum, depending on the location of the smelter. Smelters tend to be situated where electric power is both plentiful and inexpensive, such as South Africa, the South Island of New Zealand, Australia, the People's Republic of China, the Middle East, Russia, Quebec and British Columbia in Canada, and Iceland.

Over the last 50 years, Australia has become a major producer of bauxite ore and a major producer and exporter of alumina. Australia produced 62 million metric tons of bauxite in 2005. The

Australian deposits have some refining problems, some being high in silica but have the advantage of being shallow and relatively easy to mine.

Applications

General Uses

Relatively pure aluminum is prepared only when corrosion resistance or workability is more important than strength or hardness. This metal readily forms alloys with many elements such as copper, zinc, magnesium, manganese, and silicon. Aluminum alloys form vital components of aircraft and rockets as a result of their high strength-to-weight ratio. Today, almost all bulk metal materials that are referred to loosely as "aluminum," are actually alloys. For example, the common aluminum foils are alloys containing 92-99% aluminum.

A piece of aluminum metal about 15 centimeters long.

Some of the many uses for aluminum metal are in:

- Transportation (particularly automobiles, aircraft, trucks, railroad cars, marine vessels, and bicycles).
- Packaging (such as cans and foil).
- Optical coatings and mirrors, in which a thin layer of aluminum is deposited on a flat surface.
- Water treatment.
- Treatment against fish parasites such as Gyrodactylus salaris.
- Construction (windows, doors, siding, building wire, etc).
- Cooking utensils.
- Electrical transmission lines for power distribution.
- MKM steel and Alnico magnets.
- Super purity aluminum (SPA, 99.980 percent to 99.999 percent Al), used in electronics and CDs.
- Heat sinks for electronic appliances such as transistors and CPUs.

- Powdered aluminum is used in paint, and in pyrotechnics such as solid rocket fuels and thermite.
- The blades of prop swords and knives used in stage combat.

Aluminum Compounds

- Aluminum ammonium sulfate ($[Al(NH_4)](SO_4)_2$), ammonium alum is used as a mordant, in water purification and sewage treatment, in paper production, as a food additive, and in leather tanning.
- Aluminum acetate is a salt used in solution as an astringent.
- Aluminum borate (Al_2O_3 B_2O_3) is used in the production of glass and ceramic.
- Aluminum borohydride ($Al(BH_4)_3$) is used as an additive to jet fuel.
- Aluminum chloride ($AlCl_3$) is used in paint manufacturing, in antiperspirants, in petroleum refining and in the production of synthetic rubber.
- Aluminum chlorohydride is used as an antiperspirant and in the treatment of hyperhidrosis.
- Aluminum fluorosilicate ($Al_2(SiF_6)_3$) is used in the production of synthetic gemstones, glass and ceramic.
- Aluminum hydroxide ($Al(OH)_3$) is used as an antacid, as a mordant, in water purification, in the manufacture of glass and ceramic and in the waterproofing of fabrics.
- Aluminum oxide (Al_2O_3), alumina, is found naturally as corundum (rubies and sapphires), emery, and is used in glass making. Synthetic ruby and sapphire are used in lasers for the production of coherent light.
- Aluminum phosphate ($AlPO_4$) is used in the manufacture of glass and ceramic, pulp and paper products, cosmetics, paints and varnishes and in making dental cement.
- Aluminum sulfate ($Al_2(SO_4)_3$) is used in the manufacture of paper, as a mordant, in a fire extinguisher, in water purification and sewage treatment, as a food additive, in fireproofing, and in leather tanning.
- In many vaccines, certain aluminum salts serve as an immune adjuvant (immune response booster) to allow the protein in the vaccine to achieve sufficient potency as an immune stimulant.

Aluminum Alloys in Structural Applications

Aluminum alloys with a wide range of properties are used in engineering structures. Alloy systems are classified by a number system (ANSI) or by names indicating their main alloying constituents (DIN and ISO).

Aluminum is used extensively in many places due to its high strength to weight ratio. However, a designer used to working with steel will find aluminum less well-behaved in terms of flexibility.

The problems may often be addressed by redesigning parts dimensionally specifically to address issues of stiffness.

The strength and durability of aluminum alloys varies widely, not only as a result of the components of the specific alloy, but also as a result of heat treatments and manufacturing processes. A lack of knowledge of these aspects has from time to time led to improperly designed structures and given aluminum a bad reputation.

One important structural limitation of aluminum alloys is their fatigue strength. Unlike steels, aluminum alloys have no well-defined fatigue limit, meaning that fatigue failure will eventually occur under even very small cyclic loadings. This implies that engineers must assess these loads and design for a fixed life rather than an infinite life.

Another important property of aluminum alloys is their sensitivity to heat. Workshop procedures involving heating are complicated by the fact that aluminum, unlike steel, will melt without first glowing red. Forming operations where a blow torch is used therefore requires some expertise, since no visual signs reveal how close the material is to melting. Aluminum alloys, like all structural alloys, also are subject to internal stresses following heating operations such as welding and casting. The problem with aluminum alloys in this regard is their low melting point, which make them more susceptible to distortions from thermally induced stress relief. Controlled stress relief can be done during manufacturing by heat-treating the parts in an oven, followed by gradual cooling - in effect annealing the stresses.

The low melting point of aluminum alloys has not precluded their use in rocketry; even for use in constructing combustion chambers where gases can reach 3500 K. The Agena upper stage engine used a regeneratively cooled aluminum design for some parts of the nozzle, including the thermally critical throat region; in fact the extremely high thermal conductivity of aluminum prevented the throat from reaching the melting point even under massive heat flux, resulting in a reliable and lightweight component.

Household Wiring

Aluminum has about 65 percent of the conductivity of copper, the traditional household wiring material. In the 1960s aluminum was considerably cheaper than copper, and so was introduced for household electrical wiring in the United States, even though many fixtures had not been designed to accept aluminum wire. However, in some cases the greater coefficient of thermal expansion of aluminum causes the wire to expand and contract relative to the dissimilar metal screw connection, eventually loosening the connection. Also, pure aluminum has a tendency to "creep" under steady sustained pressure (to a greater degree as the temperature rises), again loosening the connection. Finally, Galvanic corrosion from the dissimilar metals increased the electrical resistance of the connection.

All of this resulted in overheated and loose connections, and this in turn resulted in some fires. Builders then became wary of using the wire, and many jurisdictions outlawed its use in very small sizes, in new construction. Eventually, newer fixtures were introduced with connections designed to avoid loosening and overheating. At first they were marked "Al/Cu," but they now bear a "CO/ALR" coding. In older assemblies, workers forestall the heating problem using a properly done crimp of the aluminum wire to a short "pigtail" of copper wire. Today, new alloys, designs, and methods are used for aluminum wiring in combination with aluminum terminations.

Precautions

Aluminum is a neurotoxin that alters the function of the blood-brain barrier. It is one of the few abundant elements that appears to have no beneficial function in living cells. A small percent of people are allergic to it — they experience contact dermatitis from any form of it: an itchy rash from using styptic or antiperspirant products, digestive disorders, an inability to absorb nutrients from eating food cooked in aluminum pans, and vomiting and other symptoms of poisoning from ingesting such products as Amphojel, and Maalox (antacids). In other people, aluminum is not considered as toxic as heavy metals, but there is evidence of some toxicity if it is consumed in excessive amounts. The use of aluminum cookware, popular because of its corrosion resistance and good heat conduction, has not been shown to lead to aluminum toxicity in general. Excessive consumption of antacids containing aluminum compounds and excessive use of aluminum-containing antiperspirants are more likely causes of toxicity. In research published in the Journal of Applied Toxicology, Dr. Philippa D. Darby of the University of Reading has shown that aluminum salts increase estrogen-related gene expression in human breast cancer cells grown in the laboratory. These salts' estrogen-like effects have lead to their classification as metalloestrogens.

It has been suggested that aluminum is a cause of Alzheimer's disease, as some brain plaques have been found to contain the metal. Research in this area has been inconclusive; aluminum accumulation may be a consequence of the Alzheimer's damage, not the cause. In any event, if there is any toxicity of aluminum it must be via a very specific mechanism, since total human exposure to the element in the form of naturally occurring clay in soil and dust is enormously large over a lifetime.

Mercury applied to the surface of an aluminum alloy can damage the protective oxide surface film by forming amalgam. This may cause further corrosion and weakening of the structure. For this reason, mercury thermometers are not allowed on many airliners, as aluminum is used in many aircraft structures.

Powdered aluminum can react with Fe_2O_3 to form Fe and Al_2O_3. This mixture is known as thermite, which burns with a high energy output. Thermite can be produced inadvertently during grinding operations, but the high ignition temperature makes incidents unlikely in most workshop environments.

Aluminum and Plants

Aluminum is primary among the factors that contribute to the loss of plant production on acid soils. Although it is generally harmless to plant growth in pH-neutral soils, the concentration in acid soils of toxic Al^{3+} cations increases and disturbs root growth and function.

The adaptation of wheat to allow aluminum tolerance is such that the aluminum induces a release of organic compounds that bind to the harmful aluminum cations. Sorghum is believed to have the same tolerance mechanism. The first gene for aluminum tolerance has been identified in wheat. A group in the U.S. Department of Agriculture showed that sorghum's aluminum tolerance is controlled by a single gene, as for wheat. This is not the case in all plants.

Aluminium Alloy

An aluminum alloy is a composition consisting mainly of aluminum to which other elements have been added. The alloy is made by mixing together the elements when aluminum is molten (liquid), which cools to form a homogeneous solid solution. The other elements may make up as much as 15 percent of the alloy by mass. Added elements include iron, copper, magnesium, silicon, and zinc. The addition of elements to the aluminum gives the alloy improved strength, workability, corrosion resistance, electrical conductivity, and/or density, compared with the pure metallic element. Aluminum alloys tend to be lightweight and corrosion resistant.

Identifying Aluminum Alloys

Alloys have common names, but they may be identified using a four-digit number. The first digit of the number identifies the class or series of alloy.

1xxx - Commercially pure aluminum also has a four-digit numerical identifier. Series 1xxx alloys are made of 99 percent or higher purity aluminum.

2xxx - The principal alloying element in the 2xxx series is copper. Heat treating these alloys improves their strength. These alloys are strong and tough, but not as corrosion resistant as other aluminum alloys, so they are usually painted or coated for use. The most common aircraft alloy is 2024. Alloy 2024-T351 is among the hardest of the aluminum alloys.

3xxx - The main alloying element in this series is manganese, usually with a smaller amount of magnesium. The most popular alloy from this series is 3003, which is workable and moderately strong. 3003 is used to make cooking utensils. Alloy 3004 is one of the alloys used to make aluminum cans for beverages.

4xxx - Silicon is added to aluminum to make 4xxx alloys. This lowers the melting point of the metal without making it brittle. This series is used to make welding wire. Alloy 4043 is used to make filler alloys for welding cars and structural elements.

5xxx - The principal alloying element in the 5xxx series is magnesium. These alloys are strong, weldable, and resist marine corrosion. The 5xxx alloys are used to make pressure vessels and storage tanks and for various marine applications. Alloy 5182 is used to make the lid of aluminum beverage cans. So, aluminum cans actually consist of at least two alloys.

6xxx - Silicon and magnesium are present in 6xxx alloys. The elements combine to form magnesium silicide. These alloys are formable, weldable, and heat treatable. They have good corrosion resistance and moderate strength. The most common alloy in this series is 6061, which is used to make truck and boat frames. Extrusion products from the 6xxx series are used in architecture and to make the iPhone 6.

7xxx - Zinc is the principal alloying element in the series starting with the number 7. The resulting alloy is heat-treatable and very strong. Important alloys are 7050 and 7075, both used to construct aircraft.

8xxx - These are aluminum alloys made with other elements. Examples include 8500, 8510, and 8520.

9xxx - Presently, the series starting with the number 9 is unused.

Properties of Aluminium Alloys

- Coarse intermetallic compounds (0.5-10 μm), obtained during solidification, or while processing, include virtually insoluble compounds, Al_7Cu_2Fe, α-Al (Fe, Mn)Si, (Mn, Fe)Al_6, $FeAl_3$, and relatively soluble compounds $CuAl_2$, Mg_2Si and Al_2CuMg. These are seen as mechanical fibers due to elongated stringers in the direction of metal flow.
- Fine, submicron particles (0.05-0.5 μm) of intermetallic compounds (which contain transition metals) such as $Al_{20}Cu_2Mn_3$, $Al_{12}Mg_2Cr$ and $ZrAl_3$. These particles retard recrystalisation and grain growth in alloys.
- Fine precipitates (up to 0.01 μm) are formed during age-hardening heat treatment cycle and are responsible for strengthening and hardening.
- Grain size and shape of the grains.
- Dislocation substructure, particularly obtained during cold working.

Melting of Aluminum Alloys

Alloys of aluminum do not melt at a fixed temperature but rather over a range of temperatures dependent on their composition. For example, alloy 5456, with approximately 5% Mg alloying constituent, has a melting range of 570 to 640 °C (1055 to 1180 °F). Melting begins at the lower end of the range and is completed at the higher end. The melting ranges for many commonly used aluminum alloys are provided in table.

Table: Melting ranges of some representative aluminum alloys.

	Melting rang			Melting rang	
Alloy	°C	°F	Alloy	°C	°F
1100	640–655	1190–1215	5052	605–650	1125–1200
2024	500–653	953–1 180	6061	580–650	1080–1205
003	640–655	1190–1210	7075	475–653	890–1175

Aluminum and aluminum alloys are melted and remelted regularly as needed for the casting of ingots or billets for subseque fabricating procedures such as rolling, extruding, drawing, or forging and also for recycling. Aluminum does not ignite or catch fire as it is being melted nor does it emit smoke or toxic gases.

Mechanical Properties of Aluminum Alloys at High Temperatures

The properties of aluminum alloys are compromised at elevated temperatures well before the metal reaches its melting temperature. For most of the alloys, strengths after significant times at temperatures above 150 to 200 °C (300 to 400 °F) are lower than those at room temperature, and the amount of the strength reduction may increase with both increasing temperature and/ or increasing time at an elevated temperature. As a result, most aluminum alloys are not usually recommended for longtime service at or above these temperatures, but they are widely used in the temperature range from room temperature up to 150 to 200 °C. Certain alloys specifically designed to maximize high-temperature resistance, such as those in the 2xxx aluminum-copper series, are usually chosen for applications in the higher end of this range.

Physical Properties of Aluminum Alloys

Several of the physical properties of aluminum and its alloys provide some protection when the alloys are near a fire in an adjacent structure and also lessen their increase in temperature in the early stages of a more immediate fire. Those physical properties include:

- The specific heat capacity of aluminum alloys (816 to 1050 J/kg·K, or 0.195 to 0.258 Btu/lb. · °F), which is approximately twice that of steel (377 to 502 J/kg · K, or 0.090 to 0.120 Btu/lb·°F). This means that it takes twice as much heat energy to raise the temperature of aluminum one degree as compared to a similar mass of steel. So in any fire, aluminum members would be relatively slower to heat. This advantage is retained as temperature increases, because the specific heat of aluminum alloys increases with temperature to the melting point.
- The thermal conductivity of aluminum and its alloys, which is 88 to 251 W/m · K, or 51 to 164 Btu (h · ft · F), and increases with increase in temperature. This is several times the value for steels (11 to 63 W/m · K, or 6 to 37 Btu [h · ft · °F]). Thus, heat from a localized source will be distributed along an aluminum structure in a much more efficient manner, enabling it to be radiated off and minimizing hot spots. Also, if the structure is sufficiently massive, the aluminum can act as a heat sink to slow the rate of increase of temperature in the early stages of a fire, increasing the period of serviceability. This might make the difference in prolonging structural endurance in a fire and allowing time to evacuate a burning structure.
- The reflectivity of aluminum, which is very high- 80 to 90% of incident radiation, many times that of bare steel, and reportedly 17 to 19 times greater than the usual painted steel structures. It remains very high, even at high temperatures and even for old and oxidized surfaces. Thus for bare aluminum or aluminum alloys, this high reflectivity also contributes to a slower rise in temperature and longer serviceability than for most structural steels during the early stages of a fire. Reflectivity is decreased if the aluminum surfaces are painted or become coated with soot.
- The emissivity of aluminum alloys (0.02 to 0.10 ε for most structural aluminum alloys), which is lower than that of carbon steels (0.10 to 0.80 ε) and stainless steels (0.27 ε). This also contributes to the ability of aluminum alloys to heat up more slowly than steels in the early stages of a fire, allowing more time for occupants to escape the fire. While emissivity varies greatly depending on surface quality and cleanliness, steel members may heat up approximately four times faster than comparable aluminum alloy members in a non-engulfing fire.

These physical properties are most important if the aluminum components of the structure are nearby or adjacent to the main fire in another structure, but they may also be helpful in the very early stages of a serious conflagration in the immediate structure. If the aluminum members become heavily coated with soot, the advantages offered by the physical properties of the original components are diminished or nonexistent.

The physical properties of several typical aluminum alloys and a widely used structural steel are illustrated in table.

Table: Physical properties of representative aluminum alloys and steel.

Material	Melting rang , °C	Boiling point, °C	Melting heat, kJ kg^{-1}	Specific heat J/ kg^{-1}· K^{-1}	Thermal conductivity	Emissivity, ε	Coefficient of thermal expansion, 10^{-6}· K^{-1}, 2100 °C
1050-O	645–658	2425	390	900	229	0.02–0.10	23.5
5083- O	574–683	2425	390	900	117	0.02–0.40	24.2
6005A-T5	605–655	2425	390	940	188	0.02–0.40	23.6
ASTM E24 steel	1400–1503	2860	250	420	54	0.10–0.80	13.5

Mechanical Behavior in Welds of Precipitation Hardened Aluminum Alloys

The precipitation hardening process requires that the second component in the aluminum alloy, is sufficiently soluble to allow extensive dissolution at an elevated temperature (solubilization treatment temperature) and that the solubility is considerably reduced at lower temperatures, such is the case of Al-Cu alloys.

According to figure, to induce precipitation hardening, the alloy is heated at a temperature higher than the solvus temperature to produce a homogeneous solid solution α, allowing dissolution of the second phase θ and eliminates the segregation of the alloy. Once, the solubilization temperature is reached, the alloy is cooling at high rate in order to limit the diffusion process of the atoms toward potential sites of nucleation. Finally, the supersaturated solid solution α_{ss} is heated at a temperature below the solvus temperature. At this temperature, the atoms have the ability to diffuse at short distances. Because the α_{ss} phase is not stable, the atoms of Cu diffuse at several sites of nucleation and a control precipitation could be formed. Precipitation hardening in metals is performed to produce a particulate dispersion of second phases to generate obstacles for the dislocation movement. The degree of hardening depends of the metallic system, the volume fraction and size of the particles and the interaction of the particles with dislocations. The interactions of the precipitated particles with dislocations are very important in terms of the magnitude of hardening. Some mechanisms have been established, involving particle bypassing by Orowan looping, bypass slip, or particle shearing. Figure shows the forces acting on a mobile dislocation in a stressed metal containing a dispersion of second phase particles.

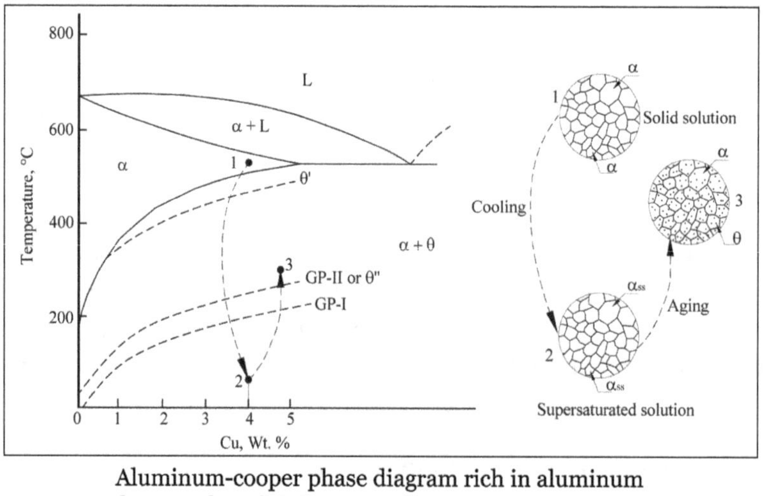

Aluminum-cooper phase diagram rich in aluminum showing the solubilization and precipitation process.

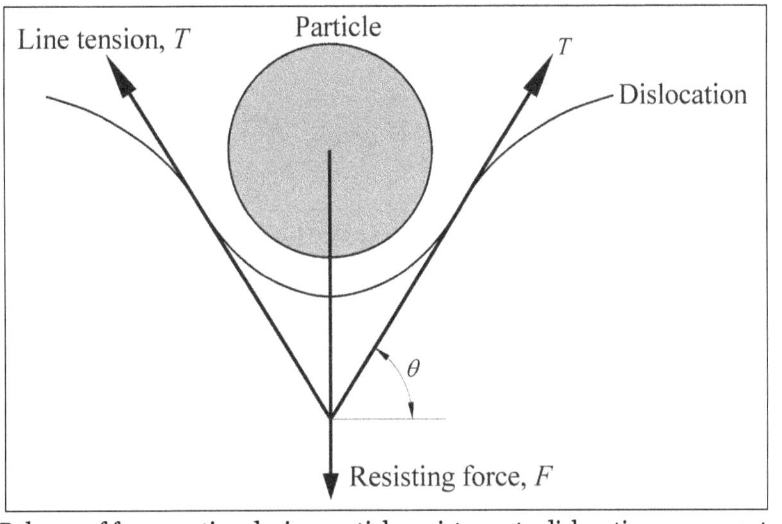

Balance of forces acting during particle resistance to dislocation movement.

Considering the equilibrium forces between the line tension T of the dislocation and the resistance force of the second phase particle F, it is obtained:

$$F = 2T \sin\theta$$

As F increases, so the bowing of the dislocations increases, i.e. θ increases. The magnitude of F is important in controlling the sequence of events. The dislocation line tension force is maximal when θ=90 °. If the particle is hard, such that F can be greater than $2T$, then dislocations will by-pass the particle either by Orowan looping or cross-slip and the particle will remain unchanged, i.e. non deformed.

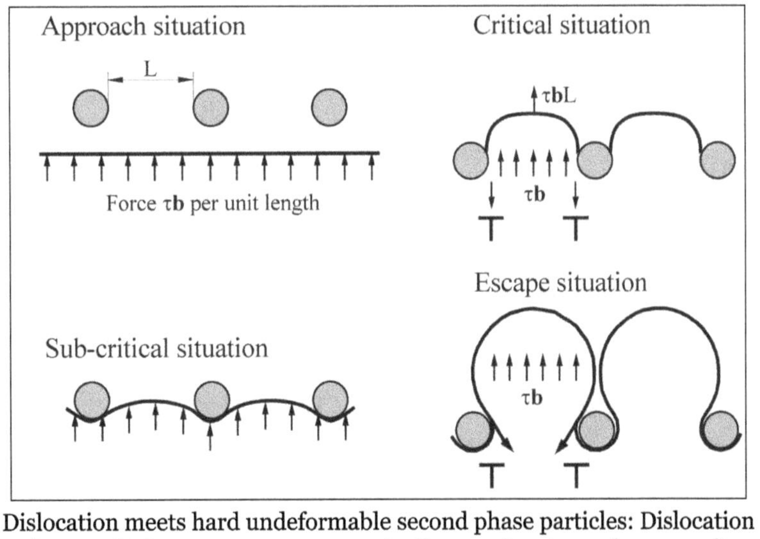

Dislocation meets hard undeformable second phase particles: Dislocation release at higher stresses may occur by Orowan looping or by cross-slip.

The actual strength of the particle under this circumstance becomes irrelevant, as the bypassing operation becomes dependent only upon the interparticle spacing. If, however, the strength of the particle is such that the maximum resistance force is attained before sin θ=1, then particles will be sheared and the dislocation will pass through the particle.

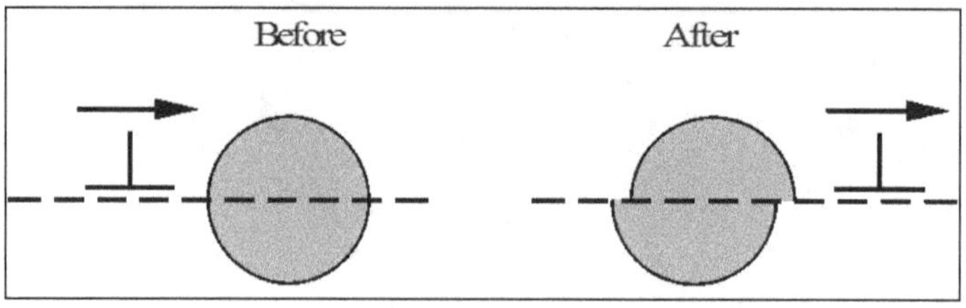

Dislocation motion may continue through second phase particles (particle cutting).

Therefore, it follows that, for a given interparticle spacing (given volume fraction and particle size), hard particles will give the maximum precipitation hardening, and this condition defines the maximum degree of hardening attainable. Soft particles give a lesser degree of hardening. Consideration of the relationship between the applied stress and the dislocation bowing, following Orowan, leads to the Orowan equation:

$$\Delta\tau_y = \frac{G\mathbf{b}}{L}$$

where $\Delta\tau_y$ is the increase in yield stress due to the particles, G the shear modulus of the matrix, b the Burgers vector of the dislocation, and L the particle spacing. The L in the Orowan equation is usually considered to be the distance between particles arranged on a square grid in the slip plane.

Ashby further developed his equation to take into account the interparticle spacing, and the effects of statistically distributed particles. The Ashby-Orowan relationship is given as:

$$\Delta\tau_y = 0.84\left(\frac{1.2G\mathbf{b}}{2\pi L}\right)\ln\left(\frac{x}{2\mathbf{b}}\right)$$

Application of the Taylor factor for polycrystalline materials, expressing the microstructural parameters in terms of the volume fraction and real diameter and converting shear stress to tensile stress, yields:

$$\Delta\sigma_y = \left(\frac{0538G\mathbf{b}f^{1/2}}{X}\right)\ln\left(\frac{X}{2\mathbf{b}}\right)$$

where $\Delta\sigma_y$ is the increase in yield strength (MPa), G is the shear modulus (MPa), b is the Burgers vector (mm), f is the volume fraction of particles, and X is the real (spatial) diameter of the particles (mm). Table I presents the hardening effect in terms of yield strength for some aluminum alloys in annealed (O) and artificial age hardened conditions (T6).

Table: Hardening increments in aluminum alloys.

	Yield strength (MPa)		
Alloy	Annealed condition (O)	Artificial age hardened condition (T6)	Difference
2014	97	417	320
6061	55	276	221
7075	103	503	400

Welding in Age Hardened Condition

The *2xxx*, *6xxx* and *7xxx* aluminum alloys are known to have a strong tendency to overage during welding, especially in fully aged condition (T6). A schematic representation of the microstructural changes in welding of aluminum alloys is shown in figure. During welding the adjacent metal to the fusion zone (welding bead) is heated and the heat affected zone (HAZ) contents two principal zones. The zone of lower temperature near to the base metal is exposed to a range of temperatures where the aging phenomena and overaging may occur. The zone of higher temperature is treated by solubilization assuming high cooling rate, the effects are less severe, because the microstructure of the material will tend to age in natural manner. However, and isothermal zone appears at within the HAZ for which the obtained temperature is located between two well defined temperatures, i.e. the artificial aging temperature of the alloy and the solvus temperature. As a result a microstructural transformation takes places due to the thermodynamic instability of the precipitates. For instance, in the case of the 6061-T6 alloy, the over-aging in the HAZ is produced by the transformation of the β'' (needle shape) precipitates into β' (rod shape) precipitates according to the following precipitation sequence:

$$\alpha_{ss} \rightarrow \text{solute clusters} \rightarrow \text{GP zones} \rightarrow \beta''(\text{needle}) \rightarrow \beta'(\text{rod}) \rightarrow \beta$$

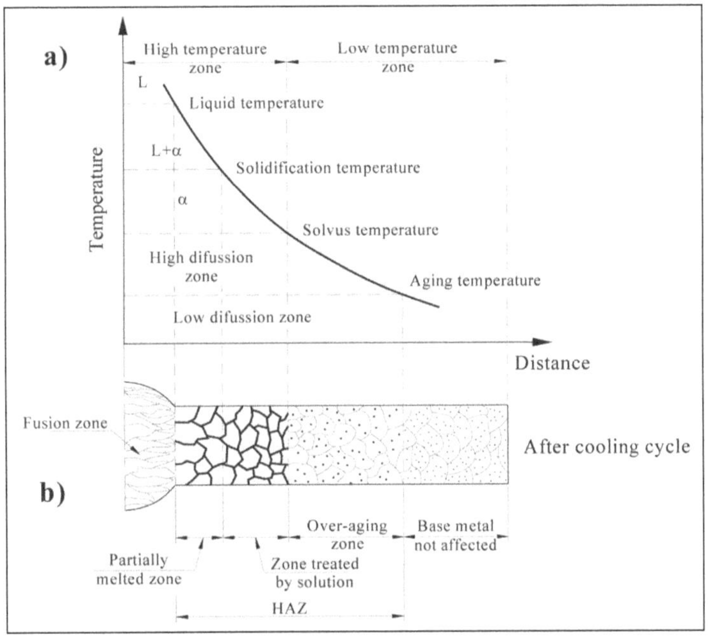

Schematic representation of the microstructural changes in heat treatable aluminum alloys during a fusion welding process, a) cooling thermal cycle from peak temperature, and b) microstructure of the welded joint at ambient temperature.

Figure, shows the weld thermal cycles for gas metal arc welding (GMAW) process in a 6061-T6 (Al-Si-Mg) alloy at different preheat conditions and their correlation between the C precipitation curve. In this sense, Myhr et al. studied the microstructural evolution during the cooling weld thermal cycle in Al-Si-Mg alloys. They determined that when the peak temperature T_p approaches 315 °C during a period of time of 10 s, the microstructure consists of a mixture of coarse rod-shape β' and fine needle-shape β'' precipitates as shown in figure and c. The transformation from β'' to β' precipitates occurs to an increasing extent with increasing peak temperatures. At T_p of 390 °C the β' is the dominant microstructural constituent, as indicated in figure.

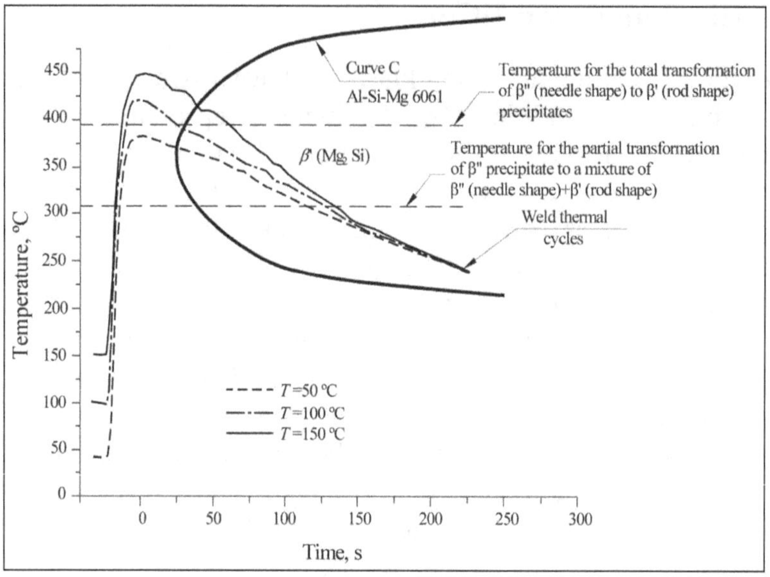

Correlation between weld thermal cycles in gas metal arc welds with three different preheating conditions, and the C precipitation curve for a 6061-T6 aluminum alloy.

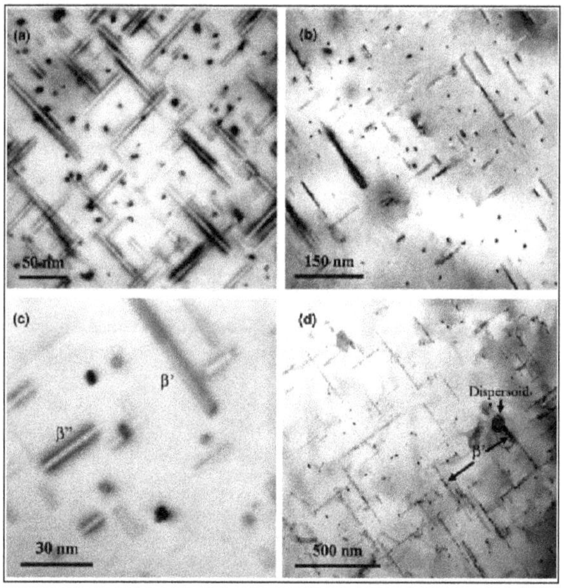

TEM bright field images of microstructures observed in the <100> Al zone axis orientation after artificial aging and Gleeble simulation (Series 1), (a) needle-shaped β" precipitates which form after artificial aging, (b) mixture of coarse rod-shaped β' particles and fine needle-shaped β" precipitates which form after subsequent thermal cycling to Tp=315 °C (10 s holding time), (c) close up of the same precipitates shown in (b) above, and (d) coarse rod-shaped β' particles which form after thermal cycling to Tp=390 °C (10 s holding time).

In addition to the microstructural transformation, after welding porosity and liquation cracking could also exist, which affect directly the mechanical behavior of the welded joints.

Porosity in welds of aluminum alloys is very complicated to control, because of the high hydrogen diffusion in liquid aluminum, the environmental interaction and the high rate of solidification.

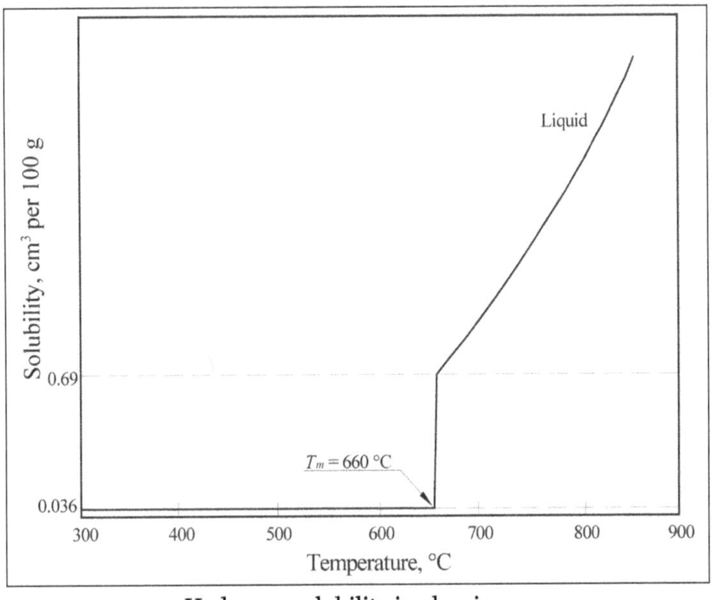

Hydrogen solubility in aluminum.

When a fusion welding process in aluminum is performed, the hydrogen diffusion in melted metal could be as high as 1.00 cm^3/g, this fact produces the formation of gas bubbles. If we consider the bubble formation in liquid metal as schematically represented in figure the bubble begins to ascend when the surface tension is overcome by the buoyant force, which tends to push the bubble to the free surface.

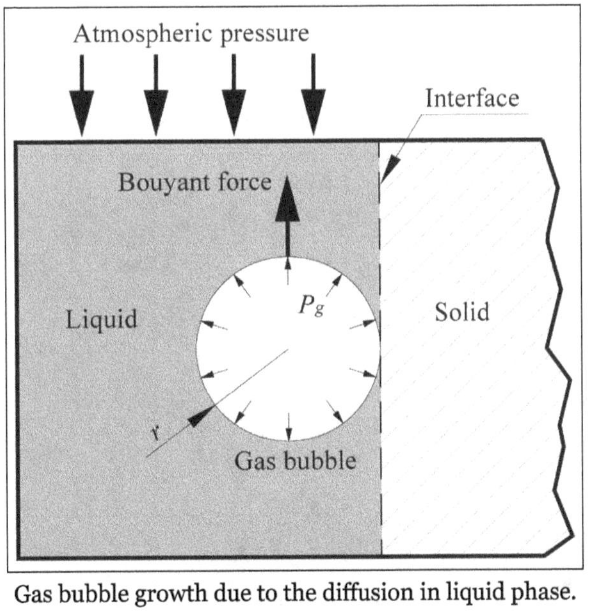

Gas bubble growth due to the diffusion in liquid phase.

The critical radius r_c for the separation of the bubble from the solid-liquid interface can be determined by means of the following expression:

$$r_c = 7.5\times10^{-4}k\left[\frac{2\sigma^*}{g(\rho_l-\rho_g)}\right]^{1/2}$$

where κ is the detaching angle of the bubble, σ* the interfacial energy between liquid and gas (~ 1 Jm^{-2} for the majority of the gas-metal systems), g the constant gravity force and ρ_l-ρ_g the difference between liquid and gas densities. Considering that densities of liquid aluminum and hydrogen at melting point of aluminum are approximately 2380 kgm^{-3} and 0.0256 kgm^{-3}, respectively, and κ=100°, the r_c is roughly 700 μm. It is to say that gas bubbles formed during welding process have to be greater than 700 μm to overcome the surface tension of the solid-liquid interface. Additionally, a low solidification rate is needed to allow the gas bubble coalescence and reach this critical value. However, in a real situation the cooling thermal cycle of the fusion zone in welding, is very fast, as demonstrated in, and porosity formation is present as shown in figure.

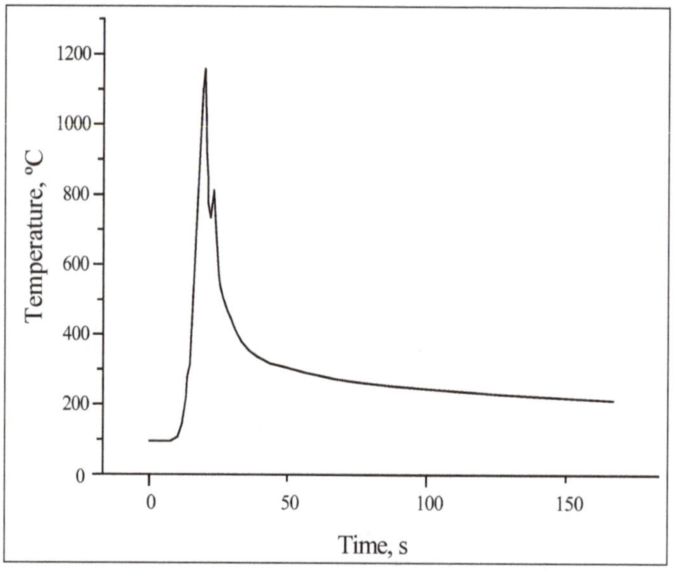

Profile of the temperature measured in the weld pool on a 6061-T6 aluminum alloy GMAW weld.

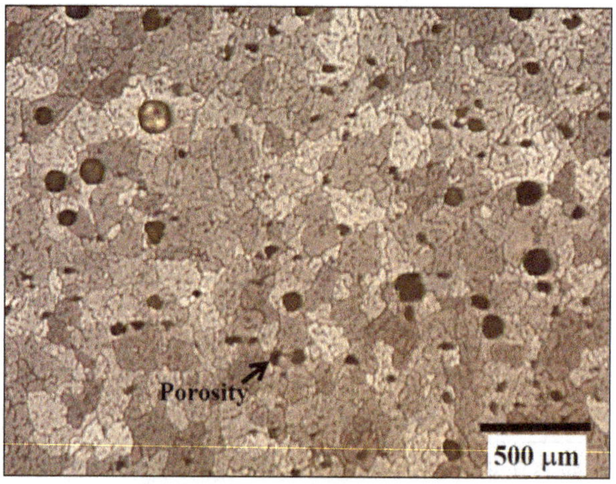

Weld pool porosity on a 6061-T6 aluminum alloy welded by GMAW.

Liquation in welds of aluminum alloys can occur in the partially melted zone (PMZ). The PMZ is the region outside the fusion zone where grain boundary liquation occurs during welding. Figure shows a portion of the PMZ in GMAW of 6061-T6 aluminum made with high silicon content filler metal (ER4043).

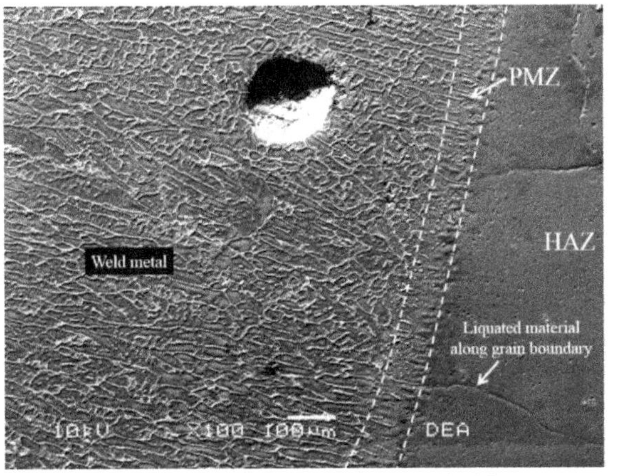

Microstructure of a 6061-T6 aluminum, welded by GMAW with ER4043 filler metal, showing the PMZ, and grain boundary liquation.

The liquation phenomenon occurs along grain boundary, although it can be presented in the grain interior. When liquation is presented, cracking can occur along grain boundary because of the tensile strains generated during welding. The weld metal composition is determined by base metal and filler metal compositions, as well as, the dilution ratio. The dilution ration is related with the amount of filler metal diluted with the base metal to form the weld metal. Metzger observed liquation cracking in full penetration, gas tungsten arc welding (GTAW) on a 6061 alloy made with Al-Mg filler metal at high dilution ratios, but not in similar welds made with Al-Si filler metals at any dilution ratio. Huang et al. have conducted studies on liquation cracking in the PMZ of full penetration welds of a 6061 alloy. They found that liquation cracking occurred in GMAW welds when an ER5356 (Al-Mg) filler metal is employed but not with an ER4043 (Al-Si).

Mechanical Behavior of Aluminum Alloys Welds

Indentation

The hardness of a material represents the plastic deformation resistance by indentation. The hardness number H, is given by the ratio between the applied load P, and a representative area A, of the residual indent:

$$H = \frac{P}{A}$$

For usual indentation, the Vickers hardness number HV, using a pyramidal square based indenter is calculated considering the true contact area. The relation of HV is given by:

$$HV = 1.8544\frac{P}{d^2}$$

where, sometimes, HV is expressed in MPa if P is given in N and d, the indent diagonal, in mm. But usually HV is given as a number and the conditions of load used.

Concerning instrumented indentation tests (IIT), which allow to plot a load-depth curve, the calculation of a hardness number can use the maximum distance (maximum indentation depth h_m, reached by the indenter during the indentation test) or the contact depth which is the indentation depth h_c, taking into account the deformation of the indent under load and calculated using the method of Olive and Pharr.

Classical indentation has been used to determine the hardness evolution in precipitation hardening aluminum alloys welds Ambriz determined Vickers microhardness profiles and mapping representation in welds of 6061-T6 aluminum alloy.

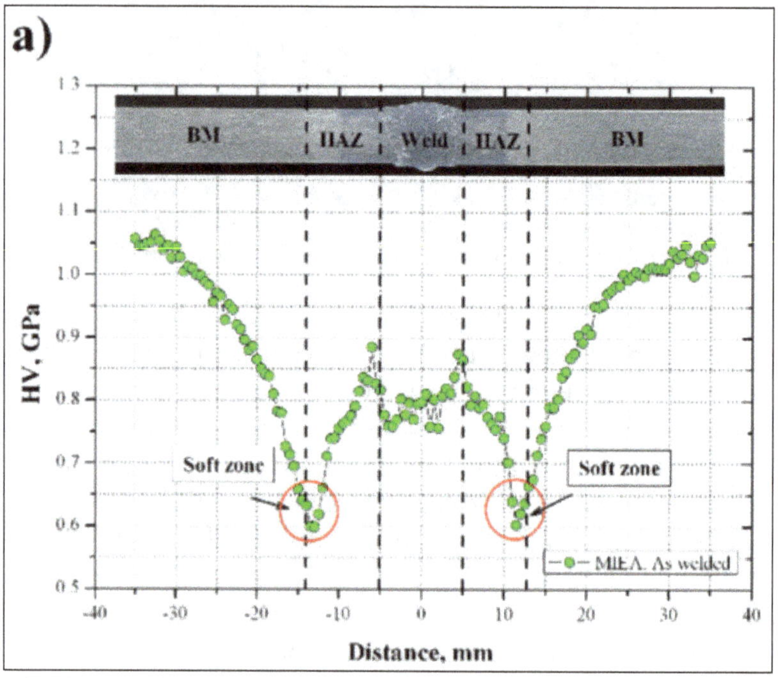

a) Vickers hardness profile determined with 1 N of applied load throughout the welded joint.

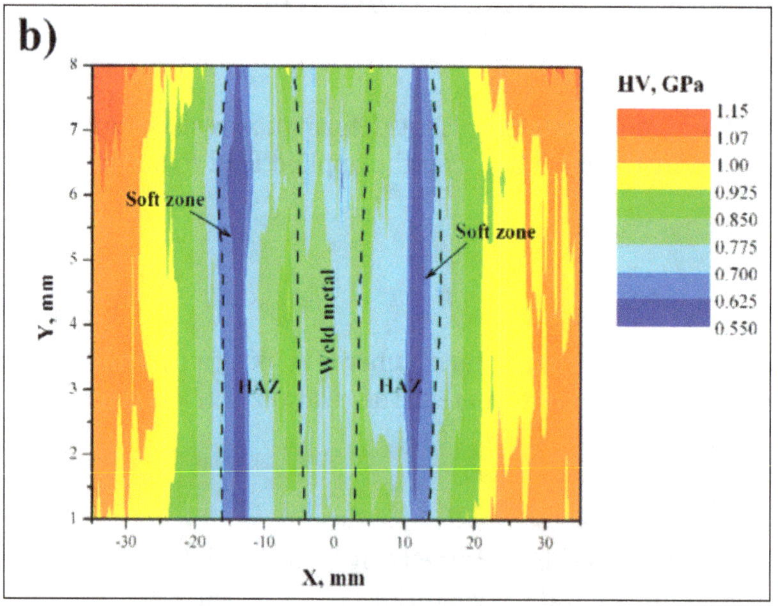

b) Vickers hardness map over the welded joint.

A significant difference for the hardness of weld metal, and HAZ with respect to base material was observed. This indicates that mechanical properties after welding will be different. It should be noted a soft zone formation in both sides of the welded joint, the hardness decrease in the soft zone is around 43% with respect to base material. This characteristic is the result of the thermodynamic instability of the β" needle-shaped precipitates (hard and fine precipitates) promoted by the high temperatures reached during the welding process. Indeed the temperatures reached during the welding process are favorable to transform the β' phase, rod-shaped according to the transformation diagram for the 6061 alloy.

Considering the hardness heterogeneity of the welded joints, instrumented indentation tests (IIT) was performed in base metal, weld metal and HAZ (soft zone). Figure, shows the evolution of the applied load as a function of indentation depth for 6061-T6 and 7075-T651 aluminum alloys welds.

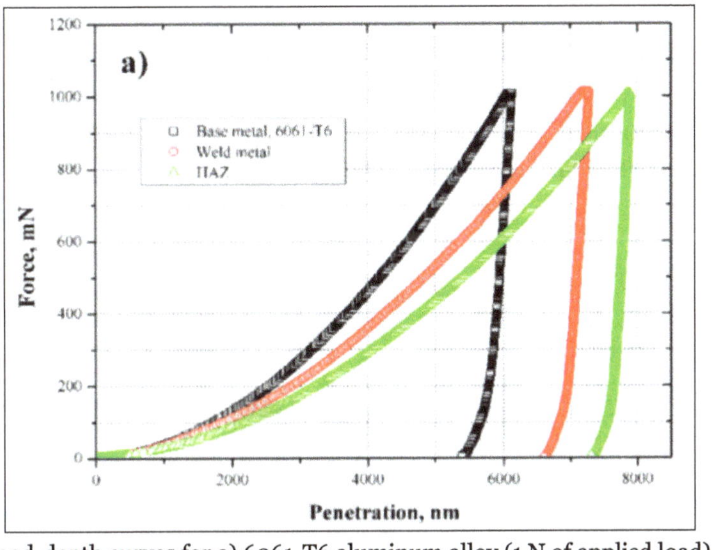

Load-depth curves for a) 6061-T6 aluminum alloy (1 N of applied load).

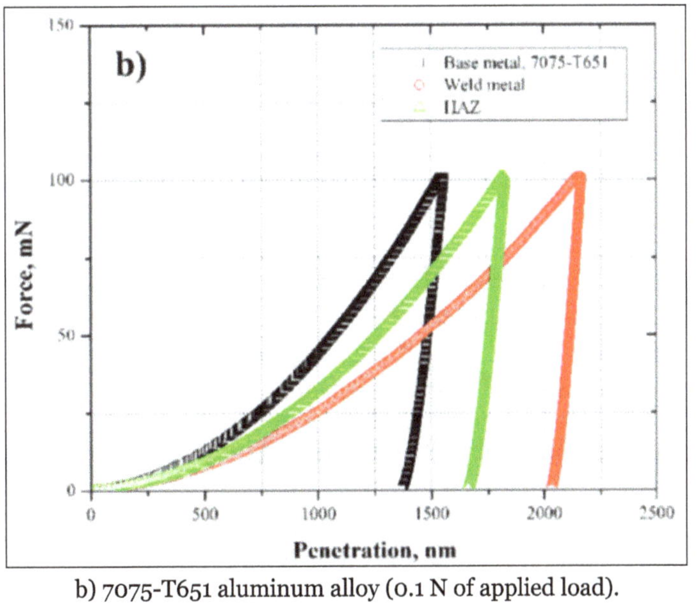

b) 7075-T651 aluminum alloy (0.1 N of applied load).

Moreover, from the instrumented indentation tests it is possible to calculate the elastic modulus which is deduced by the inverse of the unloading curve ($1/S$) as a function of the inverse of the contact indentation depth ($1/h_c$).

$$\left(\frac{dh}{dP}\right) = C_f + \sqrt{\frac{\pi}{24.5}} \cdot \frac{1}{2(\beta \cdot \gamma) \cdot E_R} \cdot \frac{1}{h_c}$$

where $\left(\frac{dh}{dP}\right)$ is the inverse of the contact stiffness, C_f the frame compliance of the indentation instrument, β geometrical factor introduce in the model of Oliver and Pharr to take into account the indenter shape and γ the factor introduced by Hay et al. in the model of Oliver and Pharr for taking into account the approximation in the Hertz's contact analysis and E_R the reduced modulus.

To calculate E_R it is needed to determine a corrective factor due to Hay. This coefficient is only dependent on the Poisson's ratio, then considering a constant value of v_m=0.3 for any region of the welded joint (base material, weld metal and HAZ), it is obtained a value of 1.067 for γ. In this condition, the slope is only linked to the reduced modulus by $0.1653/E_R$. Considering 1140 GPa and 0.07 for the elastic modulus and the Poisson's ratio for the indenter material, respectively, it is possible to determine the elastic modulus of the different zones of the welded joints, by means of the following equation:

$$E_R = \left(\frac{1-\upsilon_m^2}{E_m} + \frac{1-\upsilon_i^2}{E_i}\right)^{-1}$$

where E_m and v_m are the elastic modulus and Poisson's ratio of the material and E_i and v_i the elastic modulus and Poisson's ratio of the indenter, respectively.

Additionally, the yield strength σ_y and hardening exponent n can be obtained by indentation tests as suggested by Ambriz et al. by means of the following expression:

$$P = P_0 + 26.43\sigma_y \left\{ \frac{2}{3} + \frac{1}{2}\left(\frac{1}{3}\frac{E}{\sigma_y}\cot(\lambda)\right)^n + \frac{2}{3n}\left[\left(\frac{1}{3}\frac{E}{\sigma_y}\cot(\lambda)^n\right) - 1\right] \right\} h^2$$

where P is the applied indentation load, P_0 shift indentation load, E elastic modulus, h true indentation depth into material.

Instrumented indentation test allows to determine the properties given in table. Yield strength values determined by instrumented indentation are similar to those obtained by micro-traction test. It means that it is possible to characterize a local zone in a welded material by instrumented indentation where it is not possible by global test (tensile test).

Table: Mechanical properties obtained by instrumented indentation test for a 6061-T6 aluminum alloy welds.

Material	EIIT (GPa)	σy (MPa)	n	Po (N)
Base metal	98.6	266	0.46	0.015
Weld metal	78.0	133	0.48	0.022
HAZ	87.4	106	0.46	0.016

On the other hand, the elastic modulus obtained by instrumented indentation does not correspond with those reported in tension or compression tests for aluminum alloys (68-72 GPa) and it is not possible to establish a clear difference between weld metal and HAZ. In fact in a recent study, Chicot et al. established that the elastic modulus obtained by instrumented indentation corresponds to a bulk modulus. This was explained by the fact that for the case of indentation, the elastic and plastic strain is triaxial, whereas in tensile test, the strain used to determine the elastic modulus is uniaxial.

Tensile Properties

As possible to deduce from indentation test the tensile mechanical properties of aluminum alloys welds hardened by precipitation, are not homogeneous along the welded joint. Tensile properties in welds obtained by several welding processes of this alloys have been studied. For instance, V. Malin studied the relation between weld thermal cycles and the microstructural transformation with tensile properties of a 6061-T6 alloy welded by GMAW. tensile tests samples were taken from the welded joint as shown in figure. It is to say, a global structure effect was considered during tensile test.

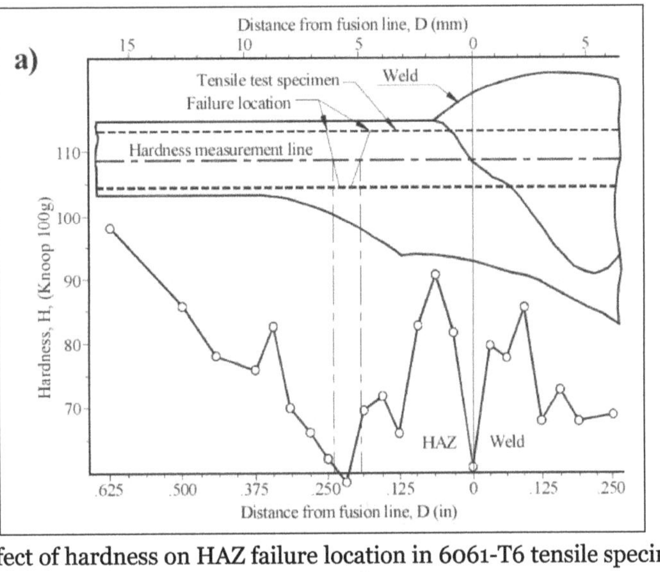

(a) Effect of hardness on HAZ failure location in 6061-T6 tensile specimen.

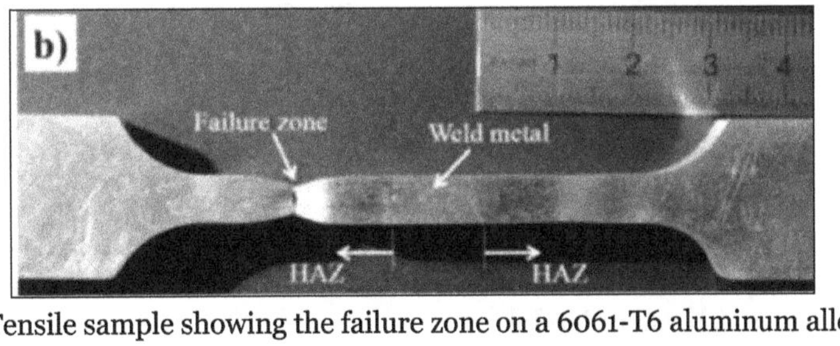

b) Tensile sample showing the failure zone on a 6061-T6 aluminum alloy.

The failure zone after tensile test was localized in the HAZ (soft zone) where the hardness of the welded joint is minimal, this result is in agreement with Ambriz.

Considering the hardness profile evolution, the true stress-strain curves in weld metal and HAZ (soft zone) of a 6061-T6 alloy welds were determined by means of micro-traction test. The individual behavior is presented in figure, as well as, its respective comparison with base metal.

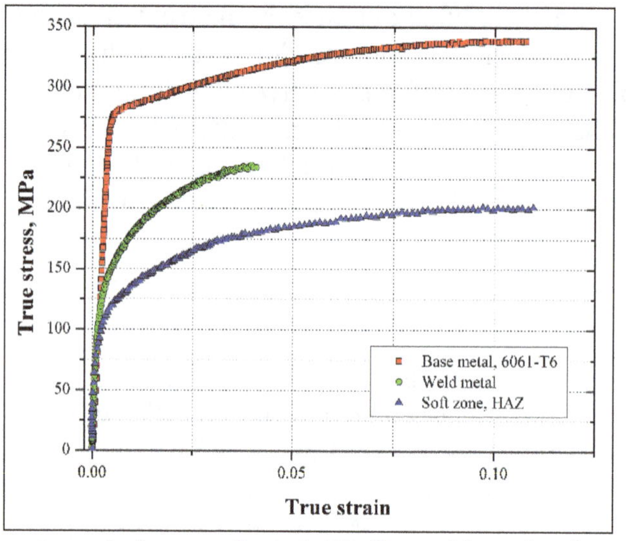

True stress-strain curves for 6061-T6 alloy, weld metal and HAZ.

The HAZ presents a reduction of the tensile strength with respect to base metal and weld metal of around 41 and 19%, respectively. This aspect was related to the over-aging phenomenon and it is explained in terms of the microstructural transformation, and precipitation sequence. Although, weld metal shows higher tensile strength than HAZ, a lower ductility is observed for weld metal. This characteristic was attribute to the porosity formation during the solidification and the high silicon content of the filler metal (ER4043) which, when mixed with the melted base metal, leads to a microstructure of eutectic silicon, which is a brittle phase that adversely affects the tensile mechanical properties of the welded joint. A summary of tensile mechanical properties for 6061-T6 and 7075-T651 aluminum alloys welds are presented in table.

Table: Tensile mechanical properties of 6061-T6 and 7075-T651 aluminum alloys welds (GMAW).

Material	E (MPa)a	σy (MPa)	σu (MPa)b	ε (%)c	H (MPa)d	n d
6061-T6	68	279	310	14.0	408	0.08
6061-T6 (weld metal, ER4043)	68	151	226	4.00	464	0.20
6061-T6 (HAZ, soft zone)	68	120	183	13.0	300	0.16
7075-T651	72	530	568	8.00	794	0.08
7075-T651 (weld joint, ER5356)	68	165	260	2.80	677	0.25

a. Best linear fitting from mean stress-strain curves; b. Obtained from $\sigma = \frac{P}{A_0}$; c. obtained from $\varepsilon = \frac{l_{f-l_0}}{l}$, d. obtained by Ramber-Osgood equation.

Fatigue of Aluminum Alloys Welds

Fatigue or fatigue damage is the consecutive modification of the materials properties with respect to the application of a cyclic stress, which can conduct to the fracture. Under uniaxial cyclic loading conditions it is possible to distinguish a stress ratio *R* determined by a maximal and minimal stress as follow:

$$R = \frac{\sigma \min}{\sigma \max}$$

As a function of σ_{max} and σ_{min}, we can obtain the constant component of the stress (amplitude stress σ_a) and the mean value of the stress (mean stress σ_m) by different loading conditions.

Some practical applications involve cyclic loading at a constant amplitude, but irregular loads as a function of time are commonly encountered. In this case, we will discuss some results in terms of a constant amplitude loading. The simplest fatigue test consists of subjecting a specimen to a cycling loading (different levels of stress amplitude σ_a), at a constant frequency and measured the number of cycles to failure N_f. The representation of stress level as a function of N_f gives the *S-N* curve or Wöhler curve. A schematic representation of a Wöhler curve is shown in figure.

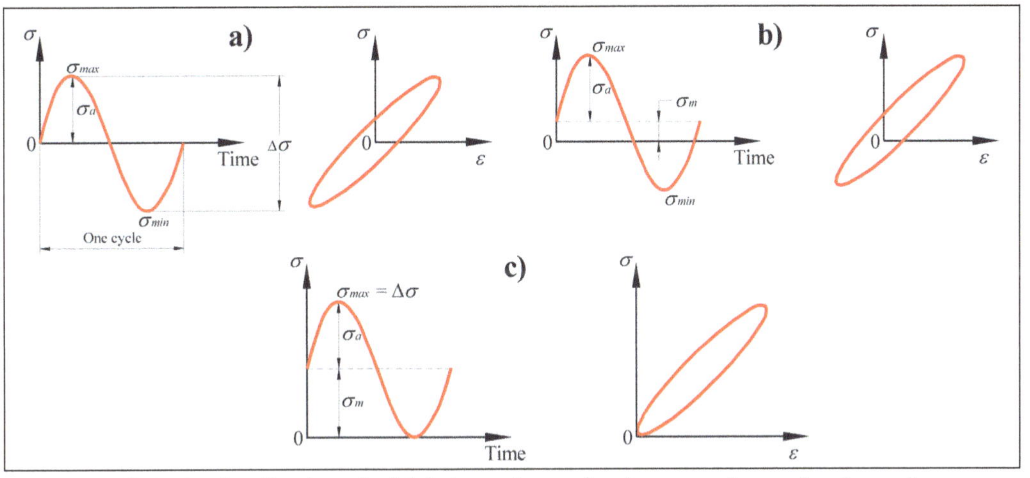

Solicitation loading in uniaxial fatigue, a) completely reversed stressing (σm=0), b) asymmetric repeated stressing (σm ≠ 0), and c) zero to tension stressing (σmin=0).

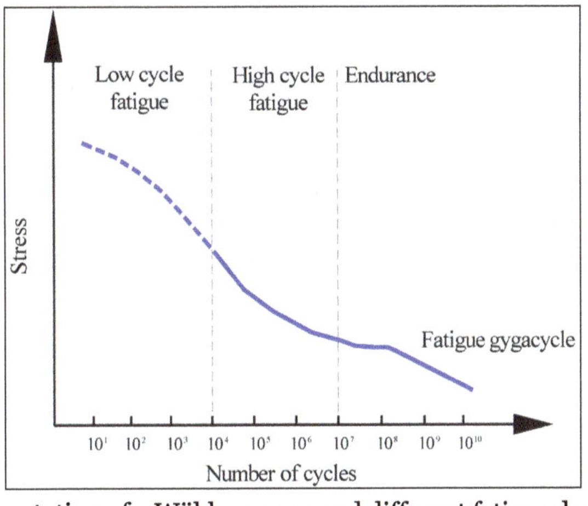

Representation of a Wöhler curve, and different fatigue domains.

In figure, it is possible to identify different domains: (i) Low cycle fatigue. In this case a high stress level is applied on the sample (normally over the yield strength of the material). Because of the high deformation during the test, the number of cycles to failure tends to be lower (10^2 to 10^4). (ii) High cycle fatigue. This is related with an elastic behavior on a macro scale of the sample, i.e. the stress level is not higher than the yield strength of the material. The failure is expected for a large number of cycles, for instance, more than 10^5. In fact the boundary between low and high cycle fatigue is not well defined by a specific number of cycles. The most important difference is that low cyclic fatigue is associated with macro-plastic deformation on each loading cycle. When the stress level in high cycle fatigue is applied, a fatigue limit or endurance limited is presented, which is represented by an asymptote in the Wöhler curve. In some metallic materials it could be obtained when the number of cycles is in the order of 10^6 to 10^7. (iii) Fatigue gigacycle. This domain corresponding with a very high number of cycles and it has been observed that fatigue limit tends to decrease when the number of cycles increases.

It is well known that fatigue damage is a surface phenomenon as indicated by Forsyth, who determined the presence of reliefs linked to the formation of localization deformation bands named persistent bands. The surface topography is traduced by the formation of intrusions and extrusions as shown schematically in figure.

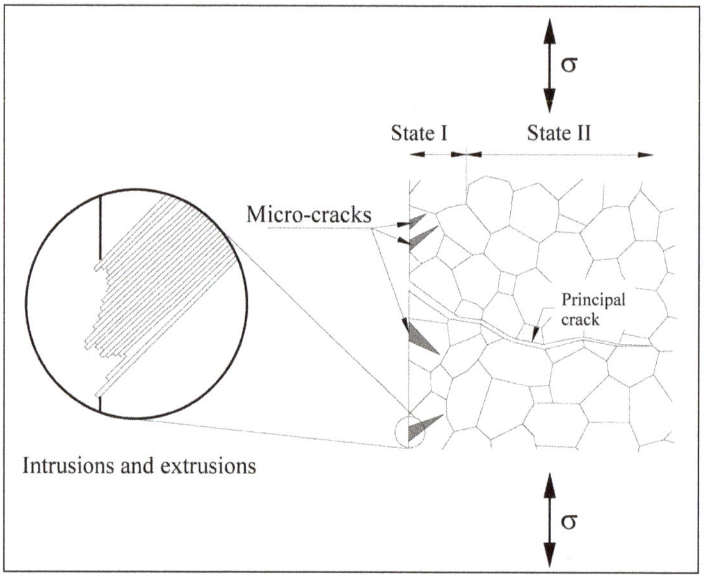

Intrusions and extrusions formation in the free surface due to the alternating slip, micro-cracks nucleation, and principal crack formation from micro-cracks.

For a uniaxial tensile test, these bands resulting in the formation of micro-cracks, which are orientated at 45 degrees with respect the traction axe. Only certain grains are affected by the formation of those bands. The persistent bands orientation and the formation of cracks on the state I, are important in the case of uniaxial and multiaxial loading. Brown and Miller introduced a useful notation in multiaxal loading for facets A and B, which are schematized in figure.

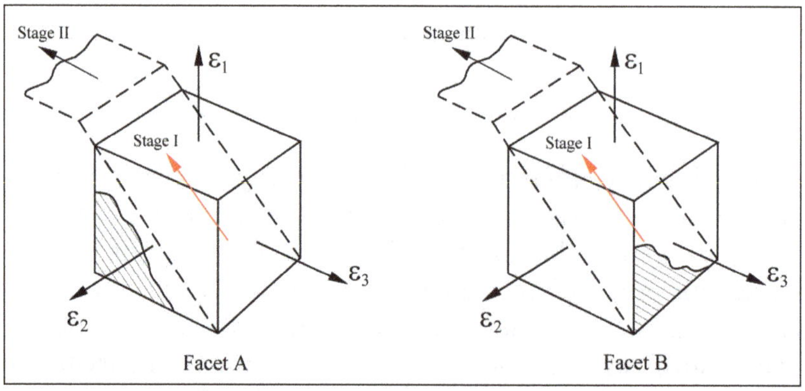

Directional aspect of the fatigue damage. Importance of the stress field orientation with respect to the plans surface, and the free surface of the material (hatch zones).

The type B facets provides a shearing vector which enters into the material, and they are more dangerous than type A facets, from which the shearing vector is tangent to the free surface of the sample. The intrusions and extrusions formation associated to the slip persistent bands, as well as, the micro-propagation of cracks in the state I are of interest at a distance of the grain size (small fatigue cracks). Thus, considering that micro-cracks are related with the crystallography aspect, once the crack encounters the first grain boundary it begins to bifurcate according to state II and the propagation at a perpendicular direction of the principal stress is obtained.

Additionally to the fatigue damage mechanism mentioned previously, in the case of welding, the stress concentration factor due to the geometry of the welding bead has a special importance. In this sense, Ambriz et al. has been quantified the effect of the welding profile generated by modified indirect electric arc (MIEA) technique on the fatigue life of a 6061-T6 aluminum alloy. In order to determine the stress concentration factor K_t in MIEA welds, a characteristic welding profile was measured as specified in figure.

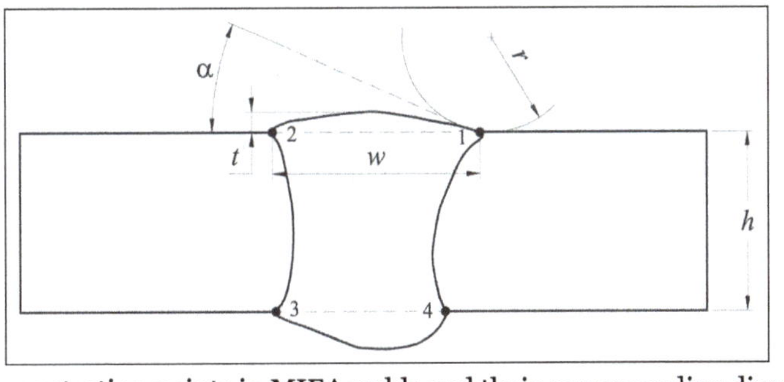

Stress concentration points in MIEA welds and their corresponding dimensions. α=angle formed by the weld reinforcement with plates, r=notched radius, t=height of the weld reinforcement, w=width of the welding profile and h=thickness of the plates.

Subsequently, uniaxial fatigue test a cyclic loading with a sinusoidal wave form at a frequency of 35 Hz and load ration R=0.1 was applied in atmospheric air at room temperature. The fatigue limit (77 MPa) was calculated employed the Locati method, and Wöhler curve was plotted between 77 and 110 MPa.

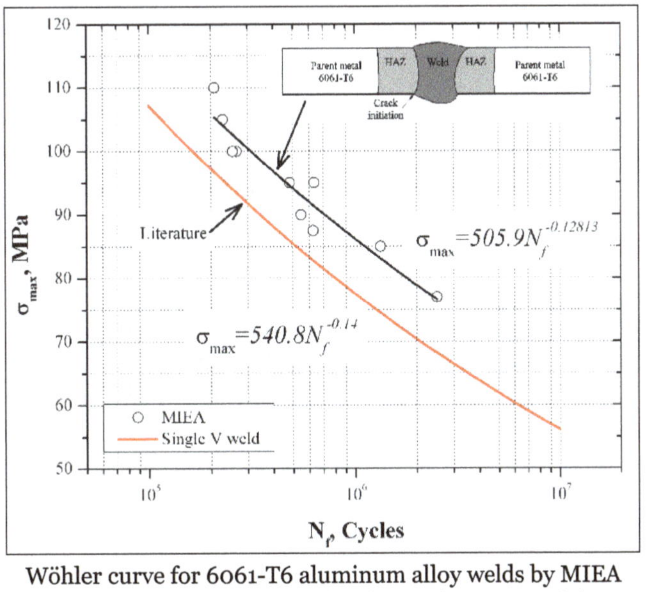

Wöhler curve for 6061-T6 aluminum alloy welds by MIEA and data found in literature for the single V groove joint.

Figure shows the maximum stress σ_{max}, as function of number of cycles to failure N_f, obtained from the fatigue test of the welded joints. The experimental results were fitted according to the following expression:

$$\sigma_{\max} = AN_f^b$$

where A, and b are experimental values determined by fitting curve.

Regarding the geometry of the welding profile, comparison of the fatigue performance exhibited by MIEA welds with the results reported in the literature for a single V joint configuration for the same aluminum alloy shows a significant improvement in fatigue life for the MIEA welded samples. The K_t effect on the crack initiation and failure is shown in figure.

Typical macrograph of the fatigue failure on MIEA welds.

Considering that, if the loading condition promotes the formation of a principal crack, it will grow according to the power law region as shown schematically in figure.

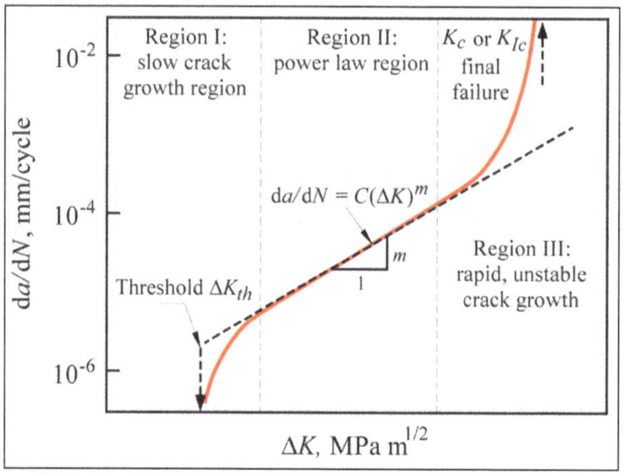

Fatigue crack growth regimes as function of ΔK.

The fatigue crack growth rate da/dN, as function of stress intensity factor range ΔK in different zones of the welded joint (base metal, weld metal and HAZ). The results were conducted on compact type specimens (CT) applying a constant amplitude cycling load with a sinusoidal wave form at a frequency of 20 Hz, a load ratio R of 0.1 and a load range of 2.5 kN in atmospheric air at room temperature. The ΔK was computed by means of the following equation:

$$\Delta K = \frac{\Delta P\left(2+\frac{a}{W}\right)}{BW^{1/2}\left(1-\frac{a}{W}\right)^{3/2}}\left[0.886+4.64\frac{a}{W}-13.32\left(\frac{a}{W}\right)^2+14.72\left(\frac{a}{W}\right)^3-5.6\left(\frac{a}{W}\right)^4\right]$$

where a is the crack length of the sample (initial crack of 8 mm), W the width for the crack propagation, and B the thickness of the sample (5 mm in this case).

Considering the stable crack growth propagation region shown in figure, the experimental results of a were plotted in da/dN versus ΔK graphs according to the following expression:

$$\frac{da}{dN} = C(\Delta K)^m$$

where C and m are constants obtained from the fitting curve. Table summarizes the best fitted values for C and m, and their correlation factor.

Table: Fitting constants obtained from experimental values.

Material	C	m	R2
6061-T6 (rolling)	5.00×10^{-7}	2.426	0.89
6061-T6 (transverse)	3.97×10^{-8}	3.320	0.97
Weld metal	2.63×10^{-14}	8.550	0.92
HAZ	3.77×10^{-9}	6.650	0.89

Figure shows the fatigue crack growth for base metal (6061-T6) in the rolling and transverse to rolling direction.

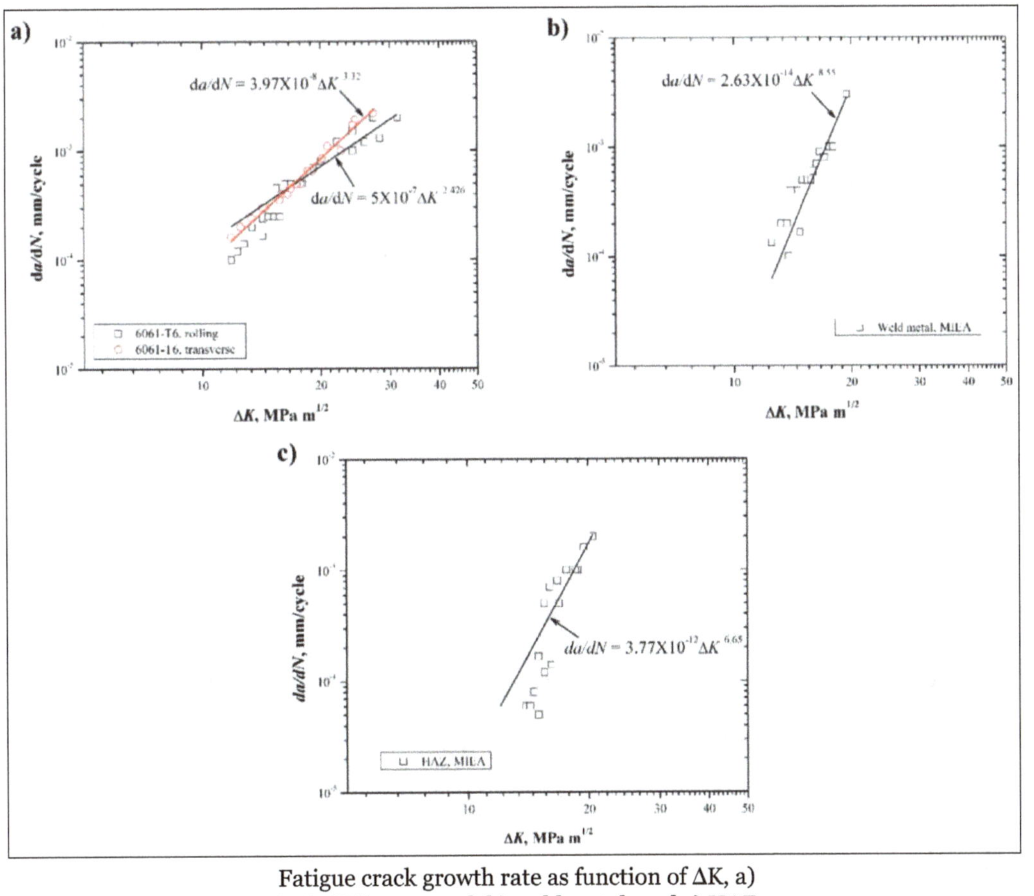

Fatigue crack growth rate as function of ΔK, a) 6061-T6 base metal, b) weld metal, and c) HAZ.

This graph shows that the microstructural characteristics (anisotropy) does not have an important influence in terms of fatigue crack growth as could be expected, taking into account that yield strength in rolling direction is higher than transverse direction. However, this is not the case for weld metal and HAZ. in which the crack tends to propagate faster than that in base metal. In the case of weld metal the faster crack growth rate in comparison with base metal is related to the low toughness due to the high silicon (~ 5.5 wt. percent) content provided by the filler metal during welding. Similarly, for the HAZ it is possible to observe that the crack growth is faster than base metal, aspect which is attribute to the microstructural transformation of fine needle shape precipitates β" into coarse bar shape precipitates β' produced by the thermal effect.

Aluminium Alloy Designation System

Most of the countries have adopted the classification adopted by International Alloy Development System (IADS). The basis of designation of wrought aluminium alloys is the four-digit system as illustrated in table. The first digit is assigned on the basis of main alloying element(s).

The second digit indicates modification of the original alloy, or impurity limits- Zero is used for original alloy, and integers 1 through 9 indicate alloy modifications. The third and fourth digits are important in the 1XXX series, and not in others alloys, i.e., in other alloys, these digits mean nothing more than a serial number. In 1XXX series, the minimum purity is denoted by these digits. For example, 1145 has a minimum purity of 99.45%, whereas as 1200 has a minimum purity of 99.00%.

Table Designation of Alloy Groups

Alloy Number	Major Alloying Elements (s)
1XXX	Commercial aluminium with minimum 99% purity
2XXX	Copper
3XXX	Manganese
4XXX	Silicon
5XXX	Magnesium
6XXX	Magnesium and Silicon
7XXX	Zinc
8XXX	Other elements such as Ni, Ti, etc.
9XXX	Unassigned

In U.K. three principal types of specifications are traditionally used for aluminium and its alloys:

- BS (British Standard)-for general engineering use.
- BS for aeronautical use (designated as the L series).
- DTD (Directorate of Technical Development) specifications are for special aeronautical applications.

Cast Aluminum Alloy Designation System

The cast alloy designation system is based on a 3 digit-plus decimal designation xxx.x (i.e. 356.0). The first digit (Xxx.x) indicates the principal alloying element, which has been added to the aluminum alloy.

Cast Aluminum Alloy Designation System

Alloy Series	Principal Alloying Element
1xx.x	99.000% minimum Aluminum
2xx.x	Copper
3xx.x	Silicon Plus Copper and/or Magnesium
4xx.x	Silicon
5xx.x	Magnesium
6xx.x	Unused Series
7xx.x	Zinc
8xx.x	Tin
9xx.x	Other Elements

Table: The second and third digits (xXX.x) are arbitrary numbers given to identify a specific alloy in the series. The number following the decimal point indicates whether the alloy is a casting (.0) or an ingot (.1 or .2). A capital letter prefix indicates a modification to a specific alloy.

Example: Alloy - A356.0 the capital A (Axxx.x) indicates a modification of alloy 356.0. The number 3 (A3xx.x) indicates that it is of the silicon plus copper and/or magnesium series. The 56 (Ax56.0) identifies the alloy within the 3xx.x series, and the .0 (Axxx.0) indicates that it is a final shape casting and not an ingot.

Heat Treatment of Aluminium Alloys

The state of the alloy in which the mechanical properties have been attained is designated by a system of temper nomenclature. It is in the form of assigning letters and digits as suffixes to the alloy number.

There are four basic tempers such as:

Letter	State of Alloy
-F	As fabricated (no guarantee of mechanical properties)
-O	Annealed. recrystallized (This is the softest temper of wrought products)
-H	Strain-hardened
-W	Solution heat-treated
-T	Heat-treated

Except for the annealed, and as fabricated tempers, the 'digit', following the letter, indicates the specific temper.

The system of designation deals separately for the non-heat treatable, strain hardening alloys on the one hand, and the heat treatable alloys on the other, such as:

H 1

Strain – Hardened only, as is indicated by the first suffix digit 1. The second digit indicates the amount of cold work (or strain hardening). The numeral '8' in H 18 indicates fully hard condition (≈ 75% reduction in area). Thus, '6' in H 16 indicates three-quarter hard; H 14 is half-hard; H 12 is quarter-hard condition. Sometimes, extra hard tempers are indicated by the numeral 9.

H 2

Strain – Hardened and partially annealed (annealing is done to reduce the harder-temper obtained by cold working to induce the strength of desired level).

H 3

Strain – Hardened and stabilised. This temper applies only to Al-Mg alloys, which have a tendency to soften with time at room temperature after cold working. Thus, to avoid this, alloys are heated (to ensure the softening process) for a short time at an elevated temperature (120° – 175 °C) to stabilise the properties. A series of H temper designations having three digits have been assigned to wrought products. For example, – Hill applies to products which have been strain-hardened less than the amount required for a controlled H 11 temper.

The heat treatable alloys are given a different nomenclature. – T applies to heat treated alloys with, or without strain hardening. The letter – T has suffixes from 2 to 10 as described below. Many times, a second digit if used indicates the amount of cold work given in between solutionised state and artificial ageing to improve the strength. For example – T 85 indicates 5% of cold work.

- T 2 – Annealed (for castings only).
- T 3 – Solution heat-treated, and then cold-worked.
- T 4 – Solution heat-treated, and naturally age-hardened.
- T 5 – Artificially aged only.
- T 6 – Solution heat-treated, and then artificially age-hardened.
- T 7 – Solution heat-treated and stabilised (by slight overaging).
- T 8 – Solution heat-treated, cold-worked, and then artificially-aged.
- T 9 – Solution heat-treated, artificially-aged, and then cold-worked.
- T 10 – Artificially-aged, and then cold-worked.

Heat Treatment Processes for Aluminium Alloys

It is essential to produce an aluminium alloy ingot of fine and uniform grain structure to obtain wrought-aluminium-alloys products with uniform and consistent properties. Even the heat treatment cycles become simpler of such alloys. Normally, the ingot is produced by direct-chill (DC), semi-continuous process.

A faster rate of solidification of the ingot results in finer dendritic arm spacings (which, thus, reduces micro-segregation in the inter-dendritic spaces), the grain size is finer and uniform, the intermetallic compounds precipitated are fine-sized. The grain size is made still smaller, if an alloy of Al-Ti, or Al- Ti-B is added to the melt before casting.

Depending on the requirement and stage of processing of aluminium alloys, the main heat treatment processes used for aluminium alloys are:

- Homogenizing.
- Annealing:
    - Stress-relieving annealing.
    - Full annealing.
- Precipitation hardening.

Homogenizing Heat Treatment

Homogenizing heat treatment is a common first heat treatment given to ingots, to:

- Reduce segregation by the process of diffusion.
- Remove low melting eutectics by the process of diffusion. Hot working can now be done without liquation, or overheating.
- Improve mechanical properties.
- Improve hot and cold workability.

Homogenising heat treatment is done at a required temperature (depending on the alloy), but normally in the range of 420 °C to 520 °C. The time at this temperature of homogenizing, though depends on the alloying elements in the alloy ingot, but is, in general related as inversely to the square of the dendritic arm spacing (in the ingot cast structure). However, the time could be 20-30 hours.

Homogenizing treatment is of greater importance for higher strength aluminium alloys. As it not only homogenizes the composition, but precipitates and redistributes the submicron intermetallic compounds of transition metals ($MnAl_6$, $ZnAl_3$, $AI_{12}Mg_2Cr$), which were supersaturated dissolved state due to faster cooling of the ingot during solidification. The uniform and finely dispersed form of these precipitates, not only controls the grain size (during heating), but also effects the mechanical properties obtained, particularly after age-hardening.

Thus, while homogenising ingots of different alloys, particularly when above mentioned precipitation occurs, the important steps to be chosen for each alloy are:

- Rate of heating to homogenising temperature.
- Temperature of homogenising.
- Time of homogenising.

Slow heating rate (75 °C/hour) helps to promote nucleation and growth of fine and uniform dispersion of the compounds. These submicron sized compounds can be further modified by controlling the rate of cooling after homogenising annealing.

Annealing

Annealing heat treatment is given to-the aluminium alloys to induce softness and ductility.

There are two types of annealing treatments given to aluminium alloys:

- Stress-relieving annealing.
- Full annealing.

Stress-relieving Annealing

It is the type of annealing in which the internal stresses developed by cold working are relieved. The main purpose of this process is to completely soften the alloy after cold working. The process consists of heating to 315 to 415 °C for a time of few minutes to 3 hours (depending on alloy and thickness). The rate of heating is not important except in alloy 3003, where rapid heating is done to prevent grain growth. It is normally advisable not to raise the temperature higher than 415 °C to avoid grain growth and oxidation.

Even intermittent annealing is also done in this range of temperature for continuing further cold work. In this process, recovery and recrystallisation occur of the cold worked alloy. The directional grains recrystallize into equiaxed grains having isotropic properties. Precautions should be taken if the alloy has been only critically deformed, because the recrystallisation results in coarse grained structure.

A fine grained equiaxed structure in semi-finished wrought products can be obtained, if recrystallisation process is interrupted just after nucleation process, or fast rate of heating is done just like heating in a salt bath (nitrate), or recrystallisation is carried out at maximum possible temperature and in minimum time.

Full Annealing

The basic aim of full annealing is to increase the softness and ductility of the aluminium alloys. This heat treatment is meant if the hardening effects of the precipitation hardening heat treatment, or of cooling from the hot working temperature, are to be removed. The full annealing heat treatment increases the softness and ductility by producing coarse and widely spaced precipitate panicles. The coarser and fewer are the intermetallic compound particles, more ductile is the alloy.

The process of full annealing consists of heating and then soaking at about 415°-440 °C for 2 hours, and then cooling slowly at 10 °C/hour up to at least 260 °C. This slow cooling also helps in growth and coalescence of the precipitating particles, so that it results in minimum hardness. In this state, the alloy can undergo high degree of deformation during cold working, as the dislocations can easily move, or cross slip, or even by pass these coarse precipitate particles.

Precipitation Hardening

Precipitation hardening is done to increase the strength and hardness of heat treatable aluminium alloys to desired values required for consistent service performance. The consistency of properties is only possible if care is taken to have a uniform microstructure in the earlier processing of the part.

The actual precipitation hardening heat treatment can be divided into three stages:

Solution Treatment

It consists of soaking the alloy at a temperature sufficiently high to achieve a homogeneous solid solution, i.e., to obtain the complete solution of all the alloying elements at a temperature within the single phase equilibrium solid solution range for the given alloy. Overheating or under-heating should be avoided.

The alloy should not be heated above solidus temperature (or initial eutectic melting temperature), as the liquation occurs of the compound at the grain boundary regions, renders the alloy brittle, and thus, adversely affects the ductility and other mechanical properties. The lower limiting temperature should be above the temperature at which complete solution occurs.

Table below gives the effect of the solutionising temperature on the strength attained after age hardening of alloy 2024-T 4 state:

Table: Effect of solutionising temperature on the strength attained after ageing in 2024-T 4.

Solutionising Temperature, °C	UTS MN/m²	Yield strength MN/m²
488	419.5	255.3
490	422.3	258.7
493	433.3	269.1
496	440.9	271.2

Table above gives solutionising temperature of some wrought aluminium alloys, and table gives solutionising temperature of cast aluminium alloys. As the soaking temperature is raised, the rate of dissolution of alloying elements into solid solution increases. The soaking time should be just enough to dissolve the solutes completely. The soaking time should be just minimum in thin clad sheets to avoid the diffusion of solutes from the alloy to the aluminium outside sheet.

Salt baths take less time than air-furnaces. H_2 gas dissolves by the reaction of alloy with water vapour in atmosphere of furnace to cause surface blisters. The blistering is normally due to overheating of the alloy. Overheating also causes grain growth to develop inferior properties. The surface reaction with water vapours can be decreased by adding fluoride salt in the furnace.

The soaking time, apart from the thickness and shape of the component, depends on conditions under which the alloy is cast. It is basically dependent on the coarseness, or fineness of the microstructure. Sand cast parts have coarser structure than permanent mould cast parts, and thus, former need longer soaking time (12 hours) to dissolve phases than latter parts (8 hours). The earlier thermal history of alloy also effects the dissolution of the alloying elements during solution-heat treatment.

For example, the rate of dissolution increases if the component has been given repeated solution heat treatment. Thus, the soaking time is decreased. Full-annealed components, having coarse precipitate particles, show decreased rate of dissolution of the alloying element (precipitates), thus, demand more soaking time. A microstructure having fine and uniform dispersion of precipitates is easy and quick to be solutionised in solution heat treatment.

Table: Heat treatment of some wrought aluminium alloys.

Alloy	Solution Temperature, °C	Post quench Process	Quenchant Temperature, °C	Time of Ageing hrs.	Temperature, of Ageing °C	Temper after precipitation
			Forgings			
2219	530 to 540	Stress relief	38 max	9-11	171-182	2219-T 852
6061	527 to 535	...	38 max	7-9	171-182	6061-T 6
6151	516 to 527	...	38 max	9-11	166-177	6151-T 6
7075	466 to 477	...	60-71	23-28	116-127	7025-T 6
			Other shapes			
2020	510 to 521	Stress relief	38 max	17-19	154-166	2020-T651
2024	488 to 500	1% cold work	38 max	11-13	188-193	2024-T 81
		6% cold work	38 max	7-9	188-193	2024-T 86
2219	530 to 540	1% cold work	38 max	17-19	171-182	2219-T 81
		7% cold work	38 max	23-25	157-168	2219-T 87
6061	521 to 538	...	38 max	7-9	171-182	6061-T 6
7075	460 to 471	...	38 max	23-28	171-182	7075-T 6

Table: Heat treatment of some aluminium casting alloys.

Alloy and Temper	Sand Casting				Permanent Mould Casting			
	Solution Temperature, °C	Time, hr.	Precipitation Temperature, °C	Time, hr.	Solution Temperature, °C	Time, hr.	Precipitation Temperature, °C	Time, hr.
319- T5	–	–	205	7-9	–	–	–	–
319-T 6	505	10-12	154	2-5	–	–	–	–
355-T 51	–	–	227	7-9	–	–	227	7-9
355-T 6	527	10-12	154	3-5	527	5-12	154	3-5
355-T 62	–	–	–	–	527	6-12	171	14-18
355-T 7	527	10-12	227	3-5	527	6-12	227	7-9

355-T 71	527	10-12	246	4-6	527	6-12	246	4-6
355-T 51	–	–	227	7-9	–	–	–	–
C 355-T 61	–	–	–	–	527	8-12	154	10-12
356-T 51	–	–	227	7-9	–	–	–	–
356-T 6	538	10-12	154	3-5	530	6-12	154	3-5
356-T 7	538	10-12	205	3-5	530	6-12	227	7-9
356-T 71	538	12	246	2-4	–		–	–
A 356-T 61	–	–	–	–	530	8-12	154	6-12

Quenching

To avoid precipitation detrimental to mechanical properties, or corrosion resistance, the solution treated aluminium alloy is quenched rapidly to obtain supersaturated solid solution at room temperature, by immersing in cold water particularly the thick parts. The fastest precipitation occurs in range 200° to 400 °C. Thus, the part should get completely submerged in quenchant before it attains a temperature of 410°C during cooling, to avoid precipitation during cooling.

There should be enough coolant, without raising its temperature during cooling, so that no precipitation occurs in this critical range of 200 °C to 400 °C. The quench-delay time, i.e., the maximum allowable transfer time (from furnace to quench bath) should be determined to fix cooling rate, the temperature of coolant, etc. A cooling rate of about 315 °C/sec in the critical range, or more is needed to avoid precipitation, though it depends on the alloy, size of the part, coolant, etc.

A cooling rate faster than that required to avoid precipitation, enhances unnecessary chances of distortion of the parts. Milder coolants like hot water (60- 80 °C), boiling water, water sprays, and air blasts can be used to reduce distortion.

Al-Zn-Mg alloys are quenched in salt bath at temperature, 180 °C, soaked there for some time, and then cooled to room temperature. Now-a-days organic quenchants are also being used. If an alloy is quenched slowly, the quenched-in vacancies are small, and thus, precipitation kinetics may be different. The precipitation may require more time.

Straightening, if required, of the aluminium alloy part may be clone in the as-quenched state. The precipitation characteristics at room temperature vary from alloy to alloy. Thus, in such cases, the parts may be refrigerated after solution treatment, and before the ageing is done.

Ageing

The choice of time-temperature cycles for precipitation should be carefully thought. The objective is to select a cycle that produces optimum precipitate size and distribution.

Thus, depending on the alloy, the ageing is classified as:

- Natural ageing – Ageing at room temperature.
- Artificial ageing – Ageing at temperatures higher than room temperature but commonly between 110° to 190 °C for 5 to 48 hours.

Most alloys require heating for a time interval (ageing) at one, or more elevated temperatures. When single ageing is to be done, a temperature is selected for which ageing time to develop high strengths properties is of convenient duration. At room temperature, alloy 2024-T 4 takes 4 days, but alloy 6061 T 4 takes a month. When multiple ageing is done, then the treatment may require more days at room temperature, followed by one, or two periods at elevated temperatures.

Precipitation hardening of castings should be based on preliminary tests. The foundry practice effects lots of properties. The principles and procedures for wrought and cast products are similar. Soaking times for solution treatment of castings are much longer as castings have very coarse structures. Castings have to be quenched more slowly, and thus, boiling water or milder quench may be used.

Some aluminium alloys yield high strength properties, if they are cold worked after quenching but before ageing. Thermo-mechanical processing has been developed when cold working is done at some stage during precipitation reaction. By this process, a uniform distribution of dislocation is made and this substructure is made stable by precipitating the compounds there. Al-Zn-Mg-Cu show high strength, toughness and fatigue strength due to precipitation and substructure hardening.

Additive Manufacturing of Al Alloys

In a product development context, the term Rapid Prototyping (RP) was widely used to describe technologies which created physical prototypes directly from digital data. The first methods for rapid prototyping became available in the late 1980s and were used to produce models and prototype parts. Users of RP technology have come to realize that this term is inadequate to describe the more recent applications of these technologies. The ASTM F-42 committee was recently formed to standardize Additive Manufacturing (AM) terminology and develop industry standards. According to their first standard, ASTM F2792-10, AM is defined as "The process of joining materials to make objects from 3D model data, usually layer upon layer, as opposed to subtractive manufacturing technologies". The basic principle of this technology is that a geometric model, initially generated using three-dimensional Computer Aided Design (3D CAD) system (e.g. CATIA, Pro/Engineer, SolidWorks), can be manufactured directly without the need of process planning.

There are many related terms used to describe AM and common synonyms include: additive fabrication, additive layer manufacturing, direct digital manufacturing, 3D printing and freeform fabrication. Within the last 20 years, AM has evolved from simple 3D printers used for rapid prototyping in non-structural resins to sophisticated rapid manufacturing systems that can be used to create functional parts in different engineering materials directly without the use of tooling. Most work to date has been conducted using polymer materials, but the development of AM processes such as Selective Laser Sintering/Melting, Electron Beam Melting and Laser Engineered Net Shaping enabled to build parts by using metallic materials, metal matrix composites and ceramic materials. Additive manufactured parts are now utilized in aerospace, automotive, medical fields and also in consumer products and military.

Additive manufacturing or 3D printing is receiving unprecedented attention from the mainstream media, investment community, and national governments around the world. The fastest-growing application, however, is in the actual manufacturing of parts for final products. In just 10 years, this important application has grown from almost nothing to more than 28% of the total global product and service revenues. The manufacturing of final parts, rather than prototyping, is where the manufacturing money is, and it is the most significant part of AM's future. Researchers and industry leaders in the European Union (EU) have identified AM as a key emerging technology. Teaming relationships have been formed between university, industry, and government entities within and across countries. The overall level of activity and infrastructure in the EU is greater to that of the U.S. in this key area. Several large cooperative projects have been funded, worth of millions of euros across Europe, among them AMAZE (Additive Manufacturing Aiming Towards Zero Waste & Efficient Production of High-Tech Metal Products), RAPOLAC (Rapid Production of Large Aerospace Components), Custom Fit (aiming at mass customized consumer and medical project manufacturing) E-BREAK (Engine Breakthrough Components and Subsystems), TiAl-Charger (Titanium Aluminide Turbochargers-Improved Fuel Economy, Reduced Emissions), and many others. Large aerospace companies, such as Boeing, GE Aviation, and Airbus, are hard at work qualifying AM processes and materials for flight. Boeing, for example, now has 200 different AM part numbers on 10 production platforms, including both military and commercial jets.

AM technologies have the potential to change the paradigm for manufacturing, away from mass production in large factories with dedicated tooling and with high costs, to a world of mass customization and distributed manufacture. AM can be used anywhere in the product life cycle from pre-production prototypes to full scale production, as well as for tooling applications or post production repair. AM processes are stimulating innovation in component design, enabling the manufacture of parts that cannot be made by traditional methods and are stimulating alternative business models and supply chain approaches. AM enables the manufacture of highly complex shapes with very few geometric limitations compared to traditional manufacturing processes.

This freedom of design, for example, has led to the technology being used to manufacture topologically optimized shapes with improved strength to weight ratios, a fundamental aspect in both aerospace and automotive design to reduce vehicle weight. For the bio-medical industry, AM is already leading to a revolution in customized medicine where dental implants, orthopedics, and hearing aids are manufactured to fit an individual's unique physiology.

However, despite many examples of growth and progress, many myths and misconceptions associated with the technology have developed:

- AM is a low-labor content "pushbutton" technology.

  At the moment, a lot of work occurs before and after the actual production of the parts: in fact data needs to be prepared at the front end, which may require CAD expertise, the repair of the models, and optimization of support structures. Moreover, main building parameters, such as layer thickness, temperature, scan speed, and a number of other options may need to be adjusted for a particular part or type of material.

- AM is fast.

  As the name implies, AM systems build parts by depositing, fusing, curing, or laminating consecutive layers of material. These layers are typically 0.025–0.250 mm in thickness, so parts often require thousands of layers. Therefore the preparation of large parts could take even several days.

- AM systems can produce anything.

  Most of AM systems can successfully build shapes that cannot be fabricated easily or at all using conventional methods of manufacturing. However, AM processes also have limitations. One is minimum wall thickness. Another is the requirement for supports on down-facing surfaces, which can be difficult to remove. Material that is trapped in internal channels can also be difficult or impossible to remove, and the dimension of the internal channels determines the degree of difficulty in removing material.

- AM systems and materials are inexpensive.

  Generally, industrial AM systems are more expensive than CNC machining centers, and starting materials are far more expensive (for example high quality powders, with controlled chemical composition, size and shape).

In any case, as AM continues to develop on multiple levels and in many directions, it will gain more momentum and respect as a method of manufacturing. Metal parts from some AM systems are already on par with their cast or wrought counterparts. As organizations qualify and certify these and other materials and processes, the industry will grow very large. 3D printing is not competing with conventional manufacturing techniques, but is instead complementing and hybridizing with them to make new things possible.

Additive Manufacturing Processes for Metals

Today, seven companies, all in Europe, offer systems based on powder bed fusion technology. Examples of powder bed fusion are direct metal laser sintering (DMLS) from EOS and selective laser melting (SLM) from Renishaw. The most popular metal systems use a laser to heat and melt fine particles in a powder bed. Parts obtained with these systems are impressive, especially the fine features and small internal channels that are possible.

Systems that instead use an electron beam as the energy source are much faster. In the case of some orthopedic implants, however, a rough surface is preferred. This, coupled with speed, is why many medical implant manufacturers have selected electron beam melting (EBM) systems from Arcam (Arcam is currently the only manufacturer to use electron beam energy in a powder bed). Many metal parts that are currently being made by AM can be cost prohibitive, difficult, or impossible to produce using conventional methods, such as casting or machining.

This gives AM an advantage, but only if the right types of parts are selected. Quality requirements play a role. Good feature detail and surface finish are possible and comparable to metal castings, but they do not match the surface quality of CNC machined parts. If a part can be produced conventionally at a reasonable cost and the volume is relatively high, it is often best to go that route at the present time. The metal AM systems are complemented by extensive CNC machining and other traditional processes.

Many types of metals are available on metal AM systems. Among the most popular for medical and aerospace applications is the titanium alloy Ti-6Al-4V. Other metals used are cobalt–chrome, stainless steels, tool steels, aluminium alloys such as AlSi10Mg, jewelry and dental gold alloys, nickel-based superalloys such as Inconel 625 and 718, and TiAl alloys. Aerospace-grade aluminium and other metals are in development. All of the unused metal powder can be recycled in the machines after sieving. Industry standards are becoming increasingly important as companies apply AM to the production of final products. In January 2012, ASTM International Committee F42 on Additive Manufacturing Technologies approved F2924-12 Standard Specification for Additive Manufacturing Titanium-6Aluminium-4Vanadium with Powder Bed Fusion. It is the first AM material standard by ASTM and it could give a boost to the additive manufacture of Ti-6Al-4V.

The technology classification agreed by the AM SIG (Special Interest Group) can be seen detailed in Table, which shows seven top level classifications for additive manufacturing technologies, below which there are a range of different material classifications, and discrete manufacturing technologies produced by a range of global companies.

Table: Classification of additive manufacturing processes defined by the AM SIG – adapted from ASTM AM classifications.

Classification	Technology	Description	Materials	Developers (Country)
Binder Jetting	3D Printing Ink-jetting S-Print M-Print	Creates objects by depositing a binding agent to join powdered material	Metal, Polymer, Ceramic	ExOne (US) VoxelJet (Germany) 3D Systems (US)
Direct Energy Deposition	Direct Metal Deposition Laser Deposition Laser Consolidation Electron Beam Direct Melting	Builds parts by using focused thermal energy to fuse materials as they are deposited on a substrate	Metal: powder and wire	DM3D (US) NRC-IMI (Canada) Irepa Laser (France) Trumpf (Germany) Sciaky (US)
Material Extrusion	Fused Deposition Modeling	Creates objects by dispensing material through a nozzle to build layers	Polymer	Stratasys (US) Delta micro Factory (China) 3D Systems (US)
Material Jetting	Polyject Ink-jetting Thermojet	Builds parts by depositing small droplets of build material, which are then cured by exposure to light	Photopolymer, Wax	Stratasys (US) LUXeXcel (Netherlands) 3D Systems (US)
Powder Bed Fusion	Direct Metal Laser Sintering Selective Laser Melting Electron Beam Melting Selection Laser Sintering	Creates objects by using thermal energy to fuse regions of a powder bed	Metal, Polymer, Ceramic	EOS (Germany) Renishaw (UK) Phenix Systems (France) Matsuura Machinery (Japan) ARCAM (Sweden) 3D Systems (US)
Sheet Lamination	Ultrasonic Consolidation Laminated Object Manufacture	Builds parts by trimming sheets of material and binding them together in layers	Hybrids, Metallic, Ceramic	Fabrisonic (US) CAM-LEM (US)
VAT Photopolymerisation	Stereolithography Digital Light Processing	Builds parts by using light to selectively cure layers of material in a vat of photopolymer	Photopolymer, Ceramic	3D Systems (US) EnvisionTEC (Germany) DWS Srl (Italy) Lithoz (Austria)

Direct Metal Laser Sintering — DMLS

Direct Metal Laser Sintering (DMLS), a trademark of EOS GmbH (Germany), is an Additive Manufacturing (AM) technology that creates parts in a layer-by-layer fashion directly from computer-aided design by selectively fusing and consolidating thin layers of powders with a scanning laser beam. DMLS is a net-shape process, producing parts with high accuracy and detail resolution, good surface quality and excellent mechanical properties. it has many benefits over traditional techniques and its application for manufacturing three dimensional objects represents one of the promising directions to solve challenging industrial problems. Moreover, since the components are built layer by layer, it is possible to design internal features and passages that could not be cast or otherwise machined. Complex geometries and assemblies with multiple components can be simplified to fewer parts with a more cost effective assembly. DMLS does not require special tooling like castings, so it is convenient for short production runs. Applications using this technology include direct parts for a variety of industries including automotive and other industries that use complex parts of small to medium size. For instance, concerning space applications, when the "buy to fly ratio" of a structure is below about 30%, meaning that more than 70% is machined away, additive manufacturing technologies become an alternative to conventional machining for complex metallic parts. And when direct manufacturing allows to produce in one shot spacecraft structures resulting from complex assemblies, then the offered possibilities are really huge. DMLS has also already been used to fabricate lightweight structures, similar to cellular structures. Inspiring to natural systems, a variety of synthetic cellular solids could be made including stochastic foams, consolidated powders, hollow sphere structures, honeycombs, textile laminates and lattice block or miniature truss structures. The cellular metal structures have been used in various industrial applications such as heat exchangers, in automotive and aerospace industries, thanks to their valuable characteristics as low density, high strength, good energy absorption and good thermal properties. Virtually any material can be produced by additive manufacturing, most of those by DMLS, with the exception of some intermetallic and high temperature alloys.

One of the most interesting types of these materials are aluminium alloys. It is rather easy to produce alloys with compositions suitable for casting, since they will melt easily, but recently wrought alloys have also been fabricated by this technique, and new ones are being studied. The microstructure of these alloys is peculiar, since the very fast cooling occurring after the melting induced by laser provides an ultra-fine microstructure and this is responsible for a significant increase of the mechanical properties that can be obtained with conventional casting processes followed by heat treatments. It is also possible, to increase specific properties of the alloy, to produce aluminium matrix composites (AMC). These could have applications in high-demanding components, for instance in automotive and aerospace industries, where it is necessary to improve stiffness, hardness and high temperature properties. Ceramic discontinuous reinforcements are the most suitable reinforcement that can be envisaged.

DMLS Process

Initially, all contour of the layer structure is exposed with a selected laser power and contour speed. As the diameter of the melted zone is usually larger than the laser diameter, it is necessary to compensate the dimensional error and the laser beam must be shifted by half the width from the contour to the inside, to make sure that the contour of the later part will correspond exactly to the original CAD data. This correction of the position is called Beam Offset (BO). The BO value is again defined with respect to the edge of the boundary, and if this value is higher or less than the correct value, the particles of the irradiated region may be not melted or over-melted. During hatching, the laser beam moves line after line several times to assure that the melting process can unroll completely. The distance between the lines is called hatching distance (h_d) and is set about one quarter of the laser beam.

Another important parameter that can lead to a distorted part or a process interruption is the layer thickness. If the value is too high, no optimal adhesion between the single layers can be realized because the melting depth is not high enough. Furthermore, mechanical tension can be generated through this layer which can lead to detachment of the layer below. If the selected value is too small, a tearing-off of a structure can happen during the recoating process, since the melted particles get struck between it and recoater blade.

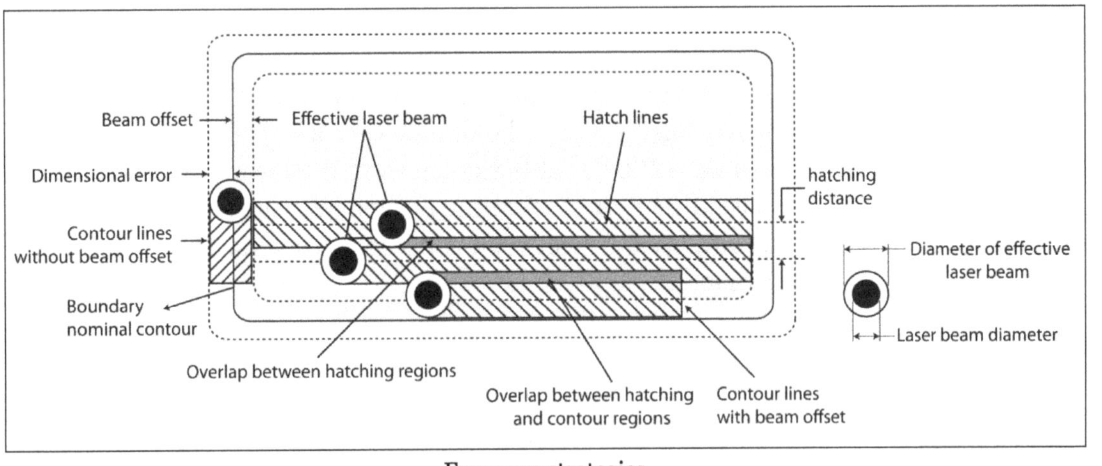

Exposure strategies.

DMLS Scanning Strategy

Many scanning options are provided in commercial SLM machines which include skywriting and hatch pattern along x, y, both in xy and alternating in xy for different requirements. DMLS EOS machine has up-skin and down-skin options to improve mechanical properties by allowing user to assign different process parameters at adjoining layers. Subsequent paragraphs illustrate salient features of scanning strategies available with DMLS process.

Skywriting

During scanning a certain time is needed to accelerate the mirrors to the desired speed. This is due to inertia of mirrors used for scanning. During this time laser beam cover some distance in which speed is not constant hence more energy is applied at the edges of the part than the inside of the part. To avoid this situation and to keep acceleration and retardation phase out of the layer geometry, skywriting option is provided in DMLS machines. If skywriting option is selected, then the mirror is accelerated already before the start of the part so that it has reached the desired speed before the beginning of exposure. Laser is switched on at the start of the part. Similarly retardation phase begins at the part end where the laser is switched off as shown in figure.

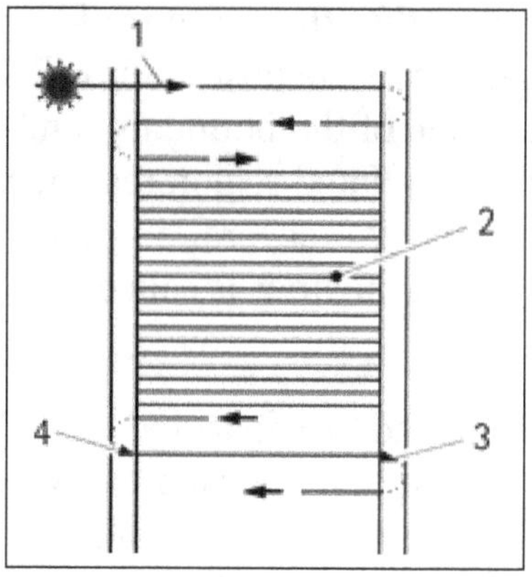

The exposure vector is shown by a solid line (1). The dashed line presents the skywriting. In this area the laser beam is run with a laser power of 0 W (3). With the beginning of the part (4), the laser power switched on with a sized value and the laser beam is running with a constant velocity.

Hatch Pattern

Four choices for hatch pattern selection are generally available, i.e., along x, along y, both in xy or alternating in xy as shown in figure. Scanning can be done either along x or along y. If both in x and y options are selected than there will be double exposure on the layer, once along x and then along y . In alternating in xy choice, direction of scanning is changed for alternating layers. Figure shows the direction of scanning rotated of 67° between consecutive layers. This is default value of the hatch pattern in DMLS EOS machine.

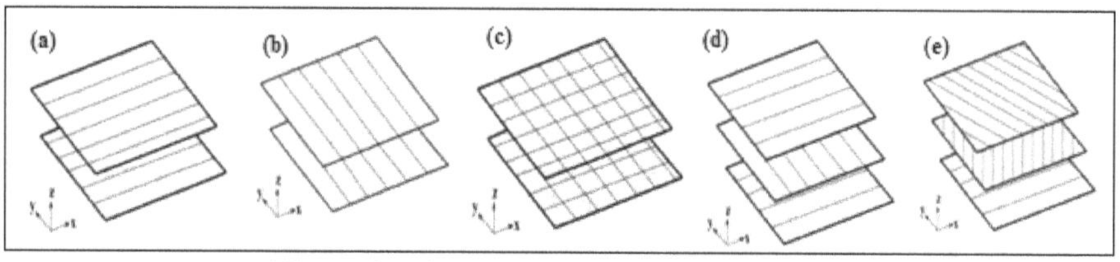

Different hatch patterns or scanning strategies.

Up-skin and Down-skin

The layer above which there is no area to be exposed, is termed up-skin, and is built in three layers. Similarly the layer below which there is no exposed area is called down-skin, and is built in two layers. The rest of the part is termed core or in-skin, and is shown in figure. Commercial machine software calculates up-skin, down-skin and core areas and different parameters can be assigned for these areas. Some overlapping area between core and up-down-skin area can also be assigned for improved joining among these zones.

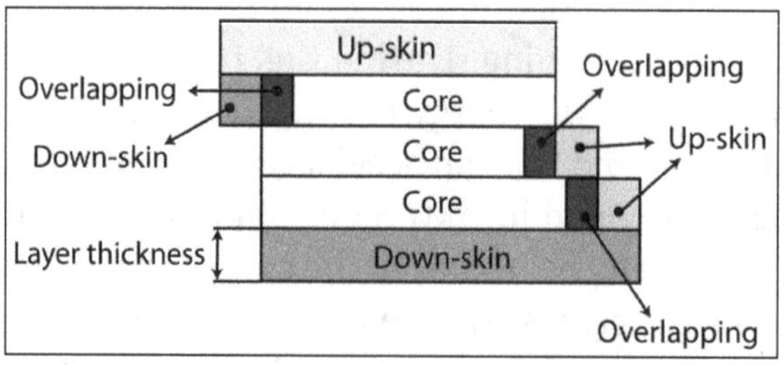

Up-skin, down-skin and core regions.

Materials Processed by DMLS

Table condenses a list of metal-based materials processed by SLM/DMLS as available in the literature.

Table: Examples of materials used in SLM/DMLS.

Family	Material	References
Metals and alloys	Steel	hot-work steel
		stainless steel 316L
		martensitic steel
		tool steel
		maraging steel
	Titanium	Ti6Al4V
		Ti6Al7Nb
	Nickel based alloy	Inconel 718
		Inconel 625
	Copper	copper
	Gold	Gold
	Aluminium	A6061
		AlSi12 and AlSi12Mg
		AlSi10Mg
Composites	MMC	Fe-graphite, Ti- graphite/diamond, Ti-SiC, AlSi-SiC, AlMg-SiC, Co-WC, Fe-SiC and Cu, Ni, Ti, C, Cu-TiC and Cu, Ni, Ti, B_4C, $Cu\text{-}TiB_2$
		Al4.5Cu3Mg-SiC
	CMC	ZrO2, Y_2O_3, Al, Al_2O_3 and TiO_2,Al,C TiC/Al_2O_3

Some of the main common features for different materials processed by SLM are summarized hereafter.

Microstructure: Small grains, non-equilibrium phases and new chemical compounds are formed during SLM of metal powders.

Mechanical properties: Due to the formation of very fine grains and non-equilibrium structures, the SLM parts could exhibit better mechanical properties (yield strength, ultimate tensile strength and ductility) than the wrought products. Yasa et al. have studied the Charpy impact test of TiAl6V4, A316L stainless steel and maraging steel parts produced by SLM. Spierings et al. have investigated the influence of three different size distribution of 316L stainless steel powder on mechanical properties and surface quality. powder with smaller size particle distribution could be easily melted and yields high density, high mechanical strength and productivity.

Anisotropy: The parts are produced by stacking layers in SLM, due to this reason the parts have some anisotropy in a particular direction. Buchbinder et al. have investigated the fatigue testing of AlSi10Mg parts produced by SLM. Their investigation concluded that sample produced in x plane exhibited higher fatigue resistance than samples produced in other orientations. Yadroitsev et al. have studied the mechanical properties of SLM parts. They found that Young's modulus of Inconel 625 parts produced in xy plane are 1.5 greater than that of parts produced in z direction. Qiu et al. studied the microstructure and tensile properties of Ti6Al4V parts produced by SLM in two different orientations. They found that Ti6Al4V parts show high tensile strength, but the Ti6Al4V samples oriented along z axis showed higher ductility than that of sample oriented in xy plane. This anisotropy was caused by the orientation of the columnar grains with respect to the tensile test direction. Thijs et al. studied about SLM of tantalum parts. Their study revealed that grains were oriented in <111> direction, because of partial re-melting of the previous layer, competitive growth mechanism and specific global direction of heat flow during SLM of tantalum. This texture formation had a large influence on yield strength during compression test of tantalum parts.

Residual stresses: These are induced in the SLM parts due to the complexity in melting behavior of layers. Re-melting and solidifying of previously melted layers can cause stress field to exceed the yield strength of the material, causing distortion of the part. Studies show that island scan strategy could reduce the residual stress of SLM parts. Shiomi et al. measured residual stress of chromium molybdenum steel parts produced by SLM. They found that: performing stress relieving treatment of build platform along with parts at about 700 °C for one hour, decreased residual stress by 70%, re-scanning of each layer reduced the residual stresses by 55% and heating of build platform to 160 °C reduced the residual stresses by 40%. Mercelis and Kruth have used crack compliance method and XRD to measure residual stress of 316L stainless steel parts produced by SLM. Zaeh and Branner have carried out Finite Element Analysis to evaluate residual stress of steels processed by SLM and they also measured residual stress by using neutron diffraction.

Surface Roughness: The surface roughness of the SLM parts is higher than that of conventionally produced parts. Vertical faces of the part have lower surface roughness than that of curved or inclined or top faces. To improve the surface finish of the SLM parts, secondary operations like shot peening, machining or abrasive jet machining are carried out. Kruth et al. showed that the surface quality of Stainless steel 316L and Ti6Al4V parts could also be improved in SLM by using Selective Laser Erosion (SLE) and re-melting. Calignano et al. have improved the roughness of AlSi10Mg parts produced by DMLS process.

Applications of Aluminium Alloy

Aerospace Alloys

Aluminium–scandium

The addition of scandium to aluminium creates nanoscale Al_3Sc precipitates which limit the excessive grain growth that occurs in the heat-affected zone of welded aluminium components. This has two beneficial effects: the precipitated Al_3Sc forms smaller crystals than are formed in other aluminium alloys and the width of precipitate-free zones that normally exist at the grain boundaries of age-hardenable aluminium alloys is reduced.Scandium is also a potent grain refiner in cast aluminium alloys, and atom for atom, the most potent strengthener in aluminium, both as a result of grain refinement and precipitation strengthening.

An added benefit of scandium additions to aluminium is that the nanoscale Al_3Sc precipitates that give the alloy its strength are coarsening resistant at relatively high temperatures (~350 °C). This is in contrast to typical commercial 2xxx and 6xxx alloys, which quickly lose their strength at temperatures above 250 °C due to rapid coarsening of their strengthening precipitates.

In principle, aluminium alloys strengthened with additions of scandium are very similar to traditional nickel-base superalloys, in that both are strengthened by coherent, coarsening resistant precipitates with an ordered $L1_2$ structure. However, Al-Sc alloys contain a much lower volume fraction of precipitates and the inter-precipitate distance is much smaller than in their nickel-base counterparts. In both cases however, the coarsening resistant precipitates allow the alloys to retain their strength at high temperatures.

The increased operating temperature of Al-Sc alloys has significant implications for energy efficient applications, particularly in the automotive industry. These alloys can provide a replacement for denser materials such as steel and titanium that are used in 250-350 °C environments, such as in or near engines. Replacement of these materials with lighter aluminium alloys leads to weight reductions which in turn leads to increased fuel efficiencies.

Additions of erbium and zirconium have been shown to increase the coarsening resistance of Al-Sc alloys to ~400 °C. This is achieved by the formation of a slow-diffusing zirconium-rich shell around scandium and erbium-rich precipitate cores, forming strengthening precipitates with composition Al_3(Sc,Zr,Er). Additional improvements in the coarsening resistance will allow these alloys to be used at increasingly higher temperatures.

Titanium alloys, which are stronger but heavier than Al-Sc alloys, are still much more widely used.

The main application of metallic scandium by weight is in aluminium-scandium alloys for minor aerospace industry components. These alloys contain between 0.1% and 0.5% (by weight) of scandium. They were used in the Russian military aircraft Mig 21 and Mig 29.

Some items of sports equipment, which rely on high performance materials, have been made with scandium-aluminium alloys, including baseball bats,lacrosse sticks, as well as bicycle frames and components, and tent poles. U.S. gunmaker Smith & Wesson produces revolvers with frames composed of scandium alloy and cylinders of titanium.

List of Aerospace Aluminium Alloys

The following aluminium alloys are commonly used in aircraft and other aerospace structures:

- 4047 aluminium,
- 6061 aluminium,
- 6063 aluminium,
- 6951 aluminium,
- 7050 aluminium,
- 7068 aluminium,
- 7075 aluminium.

The term aircraft aluminium or aerospace aluminium usually refers to 7075.

4047 alumunium is a unique alloy used in both the aerospace and automotive applications as a cladding alloy or filler material. As filler, aluminum alloy 4047 strips can be combined to intricate applications to bond two metals.

6951 is a heat treatable alloy providing additional strength to the fins while increasing sag resistance; this allows the manufacturer to reduce the gauge of the sheet and therefore reducing the weight of the formed fin. These distinctive features make aluminum alloy 6951 one of the preferred alloys for heat transfer and heat exchangers manufactured for aerospace applications.

6063 aluminium alloys are heat treatable with moderately high strength, excellent corrosion resistance and good extrudability. They are regularly used as architectural and structural members.

The following list of aluminium alloys are currently produced, but less widely used:

- 2090 aluminium,
- 2124 aluminium,
- 2324 aluminium,
- 6013 aluminium,
- 7050 aluminium,
- 7055 aluminium,
- 7150 aluminium,
- 7475 aluminium.

Marine Alloys

These alloys are used for boat building and shipbuilding, and other marine and salt-water sensitive shore applications:

- 5052 aluminium alloy,
- 5059 aluminium alloy,
- 5083 aluminium alloy,
- 5086 aluminium alloy,
- 6061 aluminium alloy,
- 6063 aluminium alloy.

4043, 5183, 6005A, 6082 also used in marine constructions and off shore applications.

Cycling Alloys

These alloys are used for cycling frames and components:

- 2014 aluminium,

- 6061 aluminium,
- 6063 aluminium,
- 7005 aluminium,
- 7075 aluminium,
- Scandium aluminium.

Automotive Alloys

6111 aluminium and 2008 aluminium alloy are extensively used for external automotive body panels, with 5083 and 5754 used for inner body panels. Bonnets have been manufactured from 2036, 6016, and 6111 alloys. Truck and trailer body panels have used 5456 aluminium.

Automobile frames often use 5182 aluminium or 5754 aluminium formed sheets, 6061 or 6063 extrusions.

Wheels have been cast from A356.0 aluminium or formed 5xxx sheet.

Cylinder blocks and crankcases are often cast made of aluminium alloys. The most popular aluminium alloys used for cylinder blocks are A356, 319 and to a minor extend 242.

Air and Gas Cylinders

6061 aluminum and 6351 aluminium are widely used in breathing gas cylinders for scuba diving and SCBA.

Aluminum Applications in the Rail Industry

Aluminum Applications and Advantages

Freight Applications: One of the most significant applications of aluminum in rail has been in the development of freight cars for the transportation of a range of commodity materials, especially coal. Typically, rail cars are designed for a 30-year life, and materials developments have tended to evolve gradually. However, much of the existing rail stock is now approaching the end of its original design life. This, together with the recent high demand for the transport of low-sulfur coal from the western mines to the eastern markets, has caused a boom in the construction of coal cars. A report forecast that deliveries of railcars will approach some 60,000 units per year by 2009 and projected a 50% growth in coal car deliveries over the five years starting in 2005. In a more recent report, AMM5 estimated that some 27% of these cars will be aluminum. The lighter aluminum railcar bodies, about two-thirds the weight of the comparable steel body, enable a greater payload. The higher payload capacity repays the higher initial cost of aluminum in less than two years, and the resistance of aluminum to corrosion by the high-sulfur coal ensures long durability for these coal cars. In addition to the operational advantages, the U.S. DOE has reported that the conversion of coal cars to aluminum has facilitated a significant decrease in the total CO_2 emissions produced by the transportation sector. To further emphasize this application, Freight Car America, a builder of rail cars, recently celebrated the delivery of the 100,000th aluminum car in a ceremony with its customer in December, 2006. Figure shows a line of open-top, hopper cars for hauling coal made using aluminum plate and extrusions.

Aluminum intensive open-top hopper cars.

The use of aluminum in coal freight cars is only one of many freight applications. Aluminum cars are used to transport a variety of metal ores and minerals from mine to market. In all these applications the high specific strength (strength to weight ratio) of aluminum enables increased payload and fuel savings. Johnstown America's website offers an excellent overview of the large number of designs available in aluminum freight cars. Recently, Freight Car America has been able to redesign its Bethgon line of cars to further increase capacity through use of alloy 6061 extrusions. In other products, aluminum sheet side walls offer improved aerodynamics and lower wind resistance and eliminate the need for paint, thereby lowering operating costs.

Aluminum Vehicle Carrier (AVC).

Another recent and popular product has been the automotive vehicle carrier. Here the car, enclosed with aluminum sheet, protects new vehicles in transit from the factory to the dealer. Extensive use of aluminum extrusions has enabled lightweight designs, with high carrying capacity for larger vans and SUVs, and especially large doors to facilitate rapid loading and unloading operations. In

Johnson America's August 3, 2004 press release on the delivery of their first production aluminum vehicle carrier (AVC) as shown in figure, Jack Thomas, president, First Union Rail was quoted as saying, "We are pleased to be innovators in the vehicle transportation field by being the first to lease the new AVC cars to Canadian National (CN). With their projected long service life and corrosion-free aluminum superstructure, we are confident the AVC cars will serve CN and its customers well." In addition, Bill Wiles, vice president, automotive for Johnstown America stated, "Rugged, yet light, the AVC is designed for minimum out of-service time, further reducing the cost of operation. A longer-term benefit is that through an AAR exemption, the aluminum superstructure never needs recertification over its extended life span".

One of the most significant advantages of aluminum is the application of internally-stiffened large hollow extrusions to fabricate moveable side walls for freight cars. As compared to conventional steel walls, this type of design enables the wall to be fabricated with many fewer parts, reducing the extent of welding required, and is significantly lighter, which has the further advantage that the wall can be manipulated by a single worker, thereby saving labor. The 6xxx series alloys (aluminum magnesium-silicon) normally used for this application offer excellent extrudability, as well as moderate strength and high corrosion resistance, and as a result the wall can be used unpainted and the cut ends of extrusions need not be sealed. The overall aluminum side wall is virtually maintenance free.

High capacity railcar with sliding aluminum sidewall.

Lastly, another advantage of aluminum freight cars is that they retain a high salvage value when recycled at the end of their life cycle. In a recent example, several hundred 100-ton hopper cars initially built around 1967 were recently sold for almost 90% of the original manufactured cost.

Commuter and Passenger Trains: The commuter and passenger rail category has an enormous variety of train designs and structures. These range from the light rail, commuter and metro trains, and trams to the more substantial inter-city and express trains. Alstom, a leading transnational corporation and producer of the European TGV and of Pendolino-type trains, has a comprehensive description of some of these trains available on their website.

In these trains the structural loads imposed by the passengers and their seating dictate the designs. Here the advantage of long aluminum extrusions has become dominant. These rail cars are made of longitudinal extrusions that run the total length of the car, and the individual extrusions are then welded together longitudinally. In the case of the Washington, DC metro system, these extrusions are 75 feet long. The overall structure is then stiffened by horizontal members that are attached around doors and windows. The floor structure is strengthened by the use of the hollow, integrally stiffened extrusions and the area near the wheel bogie is strengthened by several transverse members. This design also minimizes the number of parts required, as compared to steel fabrication, and simplifies and reduces the amount of welding. The overall result is a stiff structure that is able to handle complex loads and aerodynamic forces that result from going through tunnels, and which is about two-thirds of the weight of a comparable structure fabricated from steel.

This weight reduction is especially significant for commuter trains with frequent stops as it enables energy savings in service and reasonable acceleration between stops. The other advantages of aluminum in durability and corrosion resistance also provide low maintenance costs and long system life.

DC Metro train.

The weight reduction enabled by the use of aluminum is also critical in the newer inter-city tilting, or Pendolinotype, trains. Pendolino, which means pendulum in Italian, was the company that pioneered this tilting development and was taken over by Alstom in 2002. Just as a cyclist leans into the curve as he rounds a bend, the tilting train leans into curves and can negotiate bends at ~35% higher speed than compared to conventional trains. This development is especially important in improving the speed of inter-city service without making significant and costly track modifications or building new rail lines exclusively for high speed service. The Acela train, which operates in the northeast corridor of the U.S., is one example of the tilting train, though in this case some design considerations compromised the extent of the benefit that could be derived by the tilting mechanism.

Many tilting, Pendolino-type trains are operating throughout Europe (the French TGV), and one recently achieved the Czech Republic rail speed record of 237 km/hr. Another recent development in some of these trains is the regeneration of electricity when braking. Operational experience indicates an energy savings of ~17% through the use of regenerative braking. In all these systems, the use of aluminum technology enables extremely low axle weight and good exploitation of passenger space, all without compromising structural integrity and safety.

Acela Express train.

Advanced High Speed Trains: For the purposes of this topic, these are advanced trains that require extensively modified high speed tracks for their operation or totally new redesigned track, as is the case for the magnetic levitation (Maglev) technology. These high speed trains are under development in Germany (Transrapid) and Japan (MLX01 as a successor to the existing Shinkansen). The existing Shinkansen has developed a double decked version and increased its speed from 220 to 270 km/hr. This was feasible without overloading the wheel bogie assembly through lightweight aluminum construction technology.

Shinkansen 300 (left) and 700 Series at Tokyo Station.

The Transrapid, which uses aluminum alloys 6061 and 6063, is designed for speeds in the range of 400-500 km/hr while the Japanese MLX01 version is planned to operate at speeds of 550 km/hr. On a test track the Japanese Railway Maglev train is reported to have achieved a speed of 581 km/hr. As would be anticipated, at these design speeds, there is a premium on light weight structures. Aircraft and aerospace construction technology is increasingly being adopted. In addition to the use of longitudinal extrusions, there is considerable use of sheet material for foamed core panels for side wall and panel construction. The Maglev system closest to full scale operation and with most operational experience is in Shanghai, China. Here the system operates between the Pudong suburb and the airport and has pleased its investors with excellent high speed, low noise operation, and relatively low operating costs.

Shanghai Maglev.

While it is not clear at present what the upper limits of this advanced rail technology will be, it is obvious that both sheet and extruded aluminum and their fabrication technologies are vital to achieve these advances in transportation and infrastructure development. Table summarizes the key attributes of aluminum and their impact on the various rail applications from freight to high speed passenger systems. Clearly the multiple properties of the metal enable safe, economic, and durable performance of a host of rail systems.

Containerization

It is difficult to discuss rail transport of freight and commercial goods without at least a reference to the ubiquitous container. In the past, railroads have specified aluminum for some four out of every five containers. Goods can be placed into large containers and shipped to their destinations interchangeably by rail, road, sea, or air. The adaptability of the container for each mode of transport has simplified the transport of goods. Containers speed the loading and unloading processes, protect the goods from the elements and from pilferage, and serve as mobile warehouses. With a backbone of aluminum extrusions and with considerable use of aluminum sheet, the growth of containerization has greatly facilitated the rail transportation industry.

Aluminum Applications in Electrical Sector

Electrical Conductors

Conductors in either the 1000 or 6000 series alloys are sensible technical alternatives to copper for all electrical conductors, even in domestic wiring. A very large proportion of overhead, high voltage, power lines utilise aluminium rather than copper as the conductor on weight grounds. The relatively low strength of these grades requires that they be reinforced by including a galvanised or aluminium coated high tensile steel wire in each strand.

Aluminium alloys have a conductivity averaging 62% of the International Annealed Copper Standard (IACS) but, because of its density, it can carry more than twice as much electricity as an equivalent weight of copper.

Aluminum Applications in Building and Architecture

Aluminium is used in buildings for a wide spectrum of applications. These include roofing for factories which incorporate foil vapour barriers, windows and pre-formed sheet cladding features, doors, canopies and fronts for shops and prestigious buildings, architectural hardware and fittings, rainwater goods and replacement windows.

Aluminium structures and cladding are also used to refurbish many of the concrete structures built in the 1950-60's which are now showing signs of deterioration and spoiling.

In building applications the durability of aluminium is of paramount importance. There are a number of good examples of the durability of aluminium which may be familiar to the reader including the statue of Eros in Piccadilly Circus, London erected in 1893 and the clad dome of the church of San Gioacchino in Rome installed in 1887. More recently the oil and gas industry has employed aluminium widely in offshore structures.

The 1000, 3000, 5000 and 6000 wrought series alloys will perform, with no reduction of strength, without protection even in industrial and marine environments. They may however suffer some deterioration in their appearance and protection by painting or anodising can be advisable.

Anodised films may be clear, to preserve the 'aluminium' finish or in a limited range of colours. Painting offers a wider range of colours and an appearance similar to other painted metals, These finishing operations may also, of course, be used for purely decorative effects.

Miscellaneous Applications

High pressure gas cylinders up to fifty litre capacity for storage and transportation of CO, air, oxygen and special gases. The 6000 series alloys combine light weight, good corrosion resistance, compatibility with the product to be contained and mechanical toughness.

Machined components, the machinability of the 2000 and 6000 series alloys with additions of lead and bismuth approaches that of the free machining brasses.

Ladders and access equipment, lightweight, corrosion resistant, tough, 6000 series extrusions are used both industrially and domestically.

Sports Goods: The 2000 and 7000 series alloys are used for golf clubs and trolleys, racquets for many sports, snooker and pool cues, ski poles, often employing spin off from aerospace technology.

Road barriers and signs: Extrusions and roll formed sheet in the 6000 and 5000 series alloys provide good corrosion resistance and decorative ability.

Domestic and office furniture: The complexity and surface finish of extrusions in the 6000 series alloys coupled with the range of shapes from castings and the use of superplastically formed sheet allows designers almost unlimited scope.

Lithographic Plates: This is a high purity 1000 series sheet product which has its surface electrochemically grained then anodised to generate the base to receive the coatings used by printers.

Recrystallization Behavior of Commercial Purity Aluminium Alloys

Commercial purity aluminium alloys are largely used in the forms of foil for food packaging industries. Aluminium ingots are processed through the multi-operational steps in the consequence of foil preparation. The ingot of 7mm thickness, obtained from either direct chill (DC) casting or continuous casting process, is first cold rolled to reduce the thickness approximately 0.6 mm followed by an intermediate annealing treatment and then final roll pass is given to produce the foils of desired thickness. Therefore, the intermediate annealing treatment and consequent recrystallization behavior are prime factors for controlling microstructures as well as properties of foils. In general, the recrystallization behavior is the resultant of three steps, viz., deformation, recrystallization and grain growth. It is affected by various impurities and the precipitates forming at the operational steps.

Iron and silicon are the inevitable impurities in commercial purity aluminium alloys. The precipitation reaction due to these impurities easily occurs during heat treatment at the temperature range of 200 – 600°C, affecting cold working, softening, and corrosion resistance of the alloys. The type of intermetallic phases formed during solidification and volume fraction, amount and size of the individual particles can be controlled by changing the silicon (Si) and iron (Fe) content, i.e., by adjusting the Fe/Si ratio. The silicon influences the nature of eutectic constituent, solid solubility of other elements, formation of precipitates and dispersoids, and the transformation characteristics of the precipitates. It is necessary to control the Si content of high quality aluminium foils where material characteristics are determined by Fe/Si ratio. If the material is properly processed, then it can only be assumed that the material properties are not actually deteriorated with increased amount of Fe, Si and/or Mn. In general, composition ratio of the various alloying elements is maintained with respect to each other, but the absolute value of the single alloying element is not critical as long as primary crystallization of coarse intermetallic phases is avoided.

Plastic Deformation

Microstructure of Deformed Metals

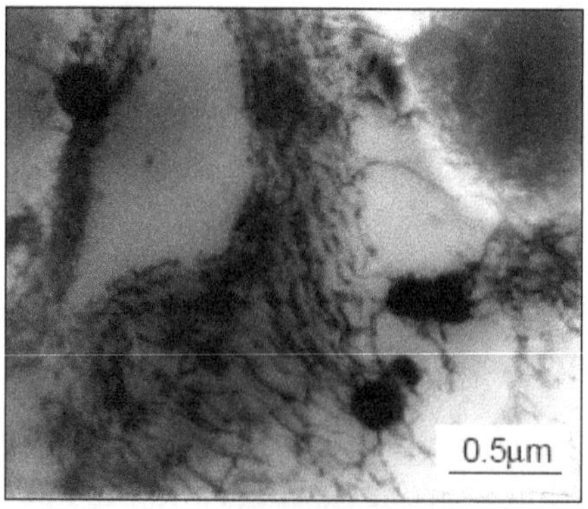

Dislocation structures of 92% cold rolled commercial purity aluminium alloy (AA1235 :0.67 Fe, 0.16Si, 0.01 Ti and rest Al).

The cold-rolled alloys are consisting high density of dislocations clearly examined through transmission electron microscopy (TEM). During deformation the dislocations are arranged as cell boundaries whereas some cell interiors are free of dislocations. Due to heavy deformation, the cell boundaries are diffuse in nature instead of simple dislocation.

However, the orientation of plastically deformed alloy changes from grain to grain or in different regions in a same grain. It happens owing to different rotations during the deformation by use of a different combination of slip systems to achieve the imposed strain. Therefore, the deformation band is nothing but a volume of constant orientation that is significantly different than the orientation of other parts of the grain, which is explained by a schematic diagram. It shows the variation of orientation at different regions of a same grain. The orientation of one part (X) of a deformed grain changes rapidly to that of a differently oriented part (Y) of the same grain across a thin boundary of a finite width called the transition bands (T) or microbands. Sometimes the orientation of deformation bands is changed twice, such as from X to Y and then Y to X. This special type of deformation band is called kink band.

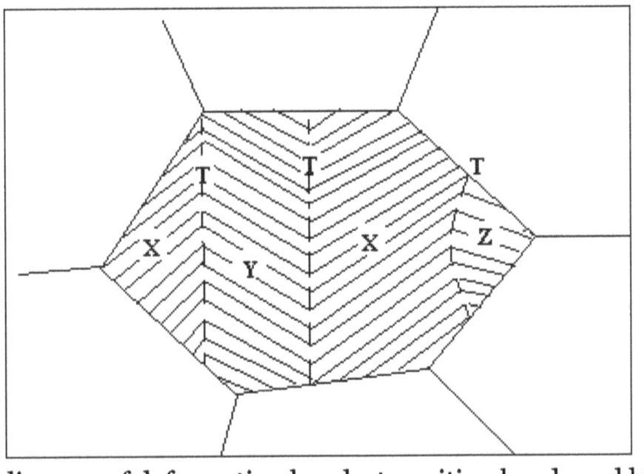

Schematic diagram of deformation bands, transition bands and kink bands.

After plastic deformation by slip mechanism, the walls of high dislocation density separate the region of low dislocation density. Such types of microstructure have been referred to as cell or subgrain. In the case of cell structures, the boundary consists of a tangled array of dislocations and appears to be a diffuse in nature. For subgrain, the boundary is sharp and consists of a well ordered dislocation array. The cell structure develops in the bulk material during or after the dislocation movement induced by the applied stress. With increasing strain, the walls become quite sharp and interiors become reasonably free of dislocations. At this stage the cell might be called subgrain.

The Effect of Particles on Deformed Microstructure

Due to rapid cooling rate in the continuous casting and direct chill casting, the alloying elements remain as supersaturated solid solution. Upon the processing operation of aluminium foils, the increased cold rolling reduction gives rise to a more pronounced dislocation entanglement. It results in the acceleration of both stored energy and precipitation. Therefore, during plastic deformation of two-phase alloys, particles affect the overall dislocation density, inhomogeneity of deformation in the matrix and deformation structure. Subsequently, the recrystallization behaviour of the alloy is also affected due to the influence of driving force as well as nucleation sites for recrystallization. If the particle is strong enough to withstand the applied stress on it, the dislocation then proceeds to encircle the particle and leaves an Orowan loop, otherwise it deforms. When the particle does not deform then extra dislocations are generated at the particle-matrix interface. On the other hand, if the particle deforms either before or after the formation of Orowan loop, no extra dislocations are generated. Generally, smaller particles are weaker than larger particles, which results in weaker slip plane. Movement of subsequent dislocations occurs on the same plane. As a consequence, slip bands are formed.

Deformation Texture

The cold rolling texture of aluminium alloy sheets has been characterized as β fiber, which is associated with plane strain deformation. The texture is varied from layer to layer of the sheet. The surface layer of the rolled sheet is known to have the shear texture, which is different from that of the plane strain deformed center layer. According to Grewen and Huber, the ideal orientations in 95% cold rolled aluminium are characterized by {112}<111>, {110}<112> and {123}<412>. The penetration depth of shear texture increased with increasing friction coefficients and rolling temperature. Above all, material parameters also influence the shear texture formation. The most important factors among material parameters are the yield strength and strain hardening exponent. The texture inhomogeneity has often been of considerable importance in rolling of pure aluminium, which has a low yield strength and work hardening exponent and tends to develop the shear texture in the surface layer. Aluminium and copper single crystals of the S-orientations develop cube texture after cold-rolling by more than 97.5% and subsequent annealing. This occurs owing to the close correlation between the formation of {001}<100> deformation structure and the development of cube texture. The cube-oriented material formed during rolling is considered to be the preferential nucleation sites of cubically aligned recrystallized grains. The factors which affect the generation of the {001}<100> deformation structure and the equivalent components during rolling are process parameters like friction between roll and sheet, rolling temperature, lubricants and the L/d value (where L and d represent the contact length and the thickness of specimens, respectively). Low viscosity lubricant such as kerosene suppresses the formation of {001}<100> deformation structure, whereas using machine oil as lubricant a cube component is formed. Thereafter, subsequent annealing treatment develops retained rolling and random texture in kerosene treated alloy and sharper cube texture in machine oil treated alloy. Truszkowski et al. and Asbeck and Mecking have reported that the shear texture is developed when $L/d \geq 5$ or $L/d \leq 0.5$.

Recovery

Subgrain Formation and its Growth

During plastic deformation of polycrystalline material (specially the alloy of medium and high stacking fault energy), unequal numbers of dislocations of two signs are generated, and the dislocations are rearranged as three-dimensional cell structure with complex dislocation tangles as cell walls. At recovery stage, the dislocations of opposite sign annihilate each other by combination of gliding and climbing mechanisms. The excess dislocations are left in the material at the end of the first stage of recovery. Upon progressing stages, these excess dislocations are arranged in a low energy configuration in the form of regular arrays or low angle grain boundaries (LAGBs). This mechanism is called polygonization, and the newly formed cells are called subgrains. In other words, the dislocation movement from the cell interiors to the cell boundary causes low angle grain boundaries, resulting in the subgrain formation. Sometimes, the dynamic recovery also helps to form a well-developed subgrain structure during deformation. In this case, recovery is the only involvement for the coarsening of subgrain structure. Although low stacking fault energy material shows poor subgrain structure, but when fine second-phase particles inhibit recrystallization then at high temperature a well-defined substructure is formed by recovery. Coarsening of substructure takes place by two methods-subgrain boundary migration and subgrain rotation.

The cell structure of a heavily cold rolled alloy is modified as a function of annealing time, examining the subgrain growth. In a commercial purity aluminium alloy (AA1235), after annealing at 250 °C for 1 h, the dislocations are examined to rearrange themselves to form a subgrain structure with an average diameter of 0.60 μm. Dislocation networks form the low angle boundary also known as subboundary. The dislocation rearrangement in the regions where the cell structure is not well-developed is described as a "disentanglement of dislocations". Consequently, the dislocation density decreases inside the subgrain and particles are precipitated at subboundary as well as inside of subgrains. After 4 h of annealing at same temperature homogeneous subgrain growth takes place with an average diameter of 2 μm. Generally, the sharp delineation of many subboundaries decreases upon annealing and ultimately fades away, in agreement with the proposed coalescence mechanism. At this stage, subgrain growth is hindered by the pinning action of Al_3Fe precipitates.

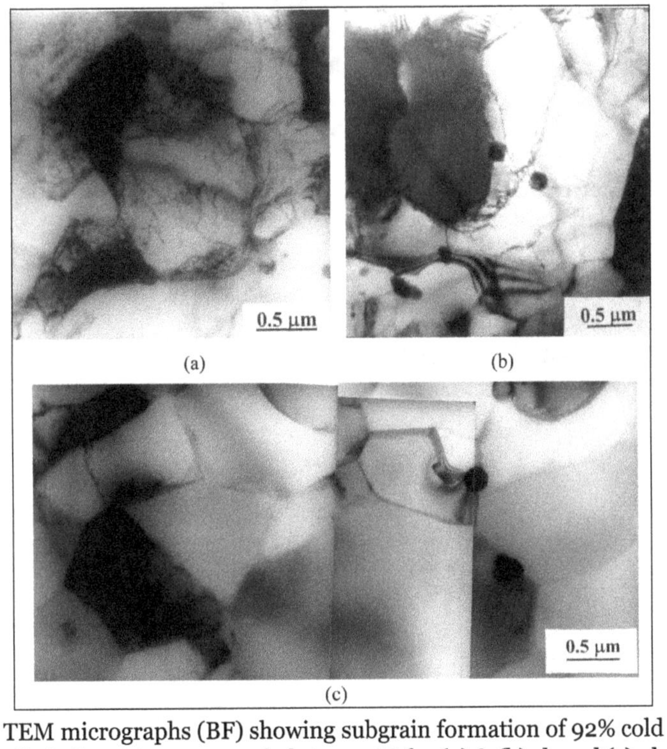

TEM micrographs (BF) showing subgrain formation of 92% cold rolled alloy AA1235 annealed at 250 °C for (a) & (b) 1h and (c) 4h.

The Effect of Second Phase Particles

In two phase alloys, second phase particles inhibit recovery by the pinning of individual dislocation or dislocations of the low angle boundaries. Therefore, the dispersion of fine particles exerts a strong pinning effect on the subgrains. Moreover, the particles stabled at high temperature improve the strength and creep properties of the alloy by pinning and stabilizing the recovered substructure. If the particles prevent the subgrain to reach the critical size for the formation of recrystallization nucleus, then recrystallization is also hindered. The evidence of dislocation pinning and restriction of subgrain growth is examined for a commercial purity aluminium alloy (AA8011) after annealing below 400 °C. Similarly, the well-defined subgrains are examined in another commercial purity aluminium alloy (AA1235) after 4h annealing at 250 °C, as shown in figure. The dislocation density change is clearly examined at some portions (X) of boundaries. The distribution of dislocation is also another observation at some points (Y). At this stage, dislocations accumulate at the grain boundary region and sometime they result in the formation of grain boundary lines at Z. It can be said that the effect of particles on subgrain growth is similar to the effect of particles on grain growth.

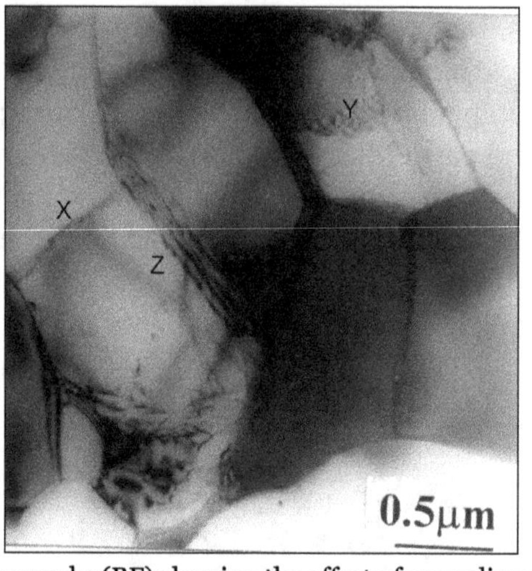

TEM micrographs (BF) showing the effect of annealing at 250 °C for 4 h on subgrain structure of 92% cold rolled bulk specimen (AA1235).

Recrystallization

The Effect of Deformation Percentage

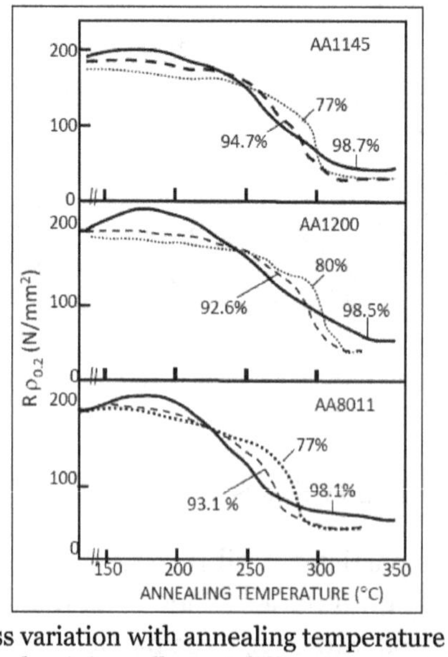

Yield stress variation with annealing temperature for three commercial purity aluminium alloys at different percentage of cold rolling.

Nucleation and growth of new recrystallized grains are closely dependent on the distribution of dislocations within the deformed grains, relating to the deformation percentages. Recrystallization does not happen below a minimum strain and the start temperature of recrystallization decreases with increasing the minimum strain level. It is known that at low strain (<5%) high activation energy for nucleation results in a low nucleation rate and a relatively high growth rate. With varying cold rolling percentage for different commercial purity aluminium alloys (AA1145, AA1200, AA8011), the degree of softening is identical for low reduction levels (<95%) and less steep for reduction levels above 95%. It corresponds to the gradual coarsening of microstructure with increasing temperature for high reduction levels (>95%), relating to the continuous recrystallization.

Nucleation of Recrystallization

The recrystallization nuclei originate at preferred sites such as prior grain boundaries, transition bands, and shear bands of deformed microstructure. The nucleation process at the grain boundary region by two different mechanisms:

- Strain induced boundary migration, where a nucleus forms when an original grain boundary bulges out.
- Subgrain coarsening by coalescence at the original grain boundary, where a nucleus forms in the consequence of subgrain growth.

With an increasing degree of deformation, the number of potential nucleation sites such as deformation bands and grain boundary bands increases significantly and the number of such sites is larger in fine grain than in coarse grain specimens. The effectiveness of nucleation sites is enhanced by the presence of intermetallic particles (Al_3Fe) with increasing degree of deformation. Potential nucleation sites are more important factor than number of nuclei. If the same numbers of nuclei form at the grain boundaries of both fine grain material and coarse grain material, then the fine grain material shows more homogeneous recrystallization. In case of heterogeneous recrystallization, all grains are not recrystallized at the same rate. It is known that crystallographic orientation affects the slip systems and strain path during deformation. Therefore, distribution and density of dislocations, large scale microstructural inhomogeneties, availability of nucleation sites, and growth rate of recrystallized grains are also dependent on crystallographic orientation.

Sequence of Precipitation and Recrystallization

It is well established that the recrystallization behavior of a deformed and supersaturated alloy is largely dependent on whether or not precipitation of the second phase particles take place simultaneously with recrystallization. Hornbogen and Köster have suggested that recrystallization occurs prior to precipitation at high temperatures, whereas precipitation takes place prior to recrystallization at low temperatures. If particles are precipitated during recrystallization they may hinder both the formation and migration of recrystallization fronts. Alternatively, if the second phase particles are precipitated in the matrix prior to cold rolling, the recrystallization behaviour will depend on the size and dispersion of the second phase particles.

The Effect of Inter Particle Spacing

At wide inter particle spacing, when only a few particles are present, there is basically no difference between the recrystallization behaviour of two-phase and single phase alloys. In this case, nucleation generally occurs at the original matrix boundaries and the ultimate recrystallized grain size may vary owing to inhomogeneous distribution of nucleation sites. With an increase in particle content of the alloy (i.e., decrease in interparticle spacing) nucleation occurs more rapidly at the lattice curvature of particle-matrix (for the particles >1 μm) interface than at the grain boundary region. In addition, due to the increase in quantity and uniform distribution of particle-matrix nucleation sites, the final recrystallized grains become more uniform and finer. However, the trend towards increased nucleation due to decrease in interparticle spacing occurs until particles maintain a critical spacing (C_1) to allow the nucleation to occur simultaneously and independently at each particle. After reaching the critical spacing (C_1), a nucleus attached to one particle would be viable itself, if it left to develop by itself. It would be non-viable if nuclei start to form simultaneously at neighbouring particles. They would then interfere with each other before reaching a viable size. Thus, due to the formation of fewer amount of viable nuclei at particle-matrix interfaces during increase in particle content (i.e., decrease in interparticle spacing) the overall nucleation rate of recrystallization decreases. Further increase in particle content leads to a second critical spacing (C_2) at which viable nuclei formation is inhibited owing to the proximity of particles. In such circumstances nucleation is likely to occur predominantly at the original matrix grain boundaries. As a consequence, the nucleation as well as recrystallization rate drastically get reduced if the inter particle spacing falls below C_2 .

The Effect of Particle Size

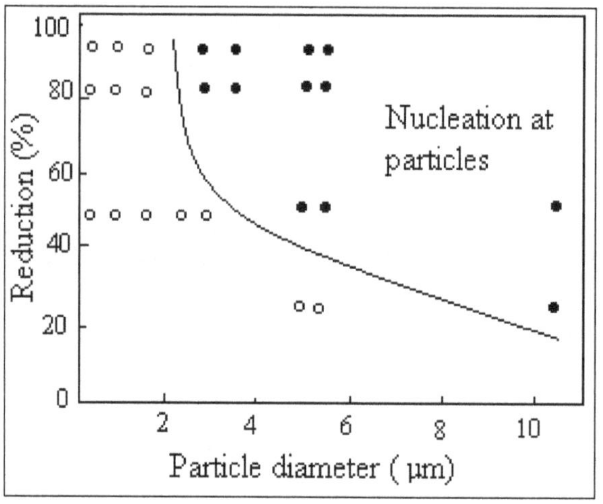

The conditions of deformation and particle size for which nucleation of recrystallized grains is observed to occur at particles.

At the time of deformation, large particles (>1 μm) lead to a heterogeneous distribution of dislocations, whereas fine particles (<0.1 μm) give rise to a homogeneous distribution of dislocations. As a consequence, the number of possible recrystallization nucleation sites increases for large particles and decreases for fine particles. This can be related to the degree of deformation, explaining the nucleation of recrystallized grains at the particles greater than 2 μm for a highly deformed metal. The fine particles also inhibit sub-boundary migration and thus the nucleation process is retarded. As the particles become more finely spaced, nearly all cell boundaries will be pinned by the particles at the end of the deformation stage. In this case, when recrystallization occurs, the mechanism of nucleation is not well understood, but the kinetics of the process is certainly very slow. Closely spaced, thermally stable particles preserve the deformed/recovered microstructure up to the melting point of the matrix. Kim et al. have revealed that the retention of dislocation substructure at high temperature provides an additional strengthening mechanism to the dispersion hardening of the alloys used at high temperature structural applications. With the decrease in the particle size and spacing between particles, recrystallization kinetics is retarded, but final grain size becomes quite large. Several researchers have studied the relationship between dispersion characteristics, deformation substructure, and recrystallization.

Type of Particles in Commercial Purity Aluminium

In commercial purity aluminium alloys, the main impurity Fe combines with both Al and Si to form a large variety of phases during solidification or during subsequent thermomechanical processing. Shoji and Fujikura have identified three types of precipitates Si, α-AlFeSi, and Al_3Fe in cold rolled commercial alloy AlFe0.6Si0.16 (all in wt%) after annealing in salt bath at different temperatures 300, 400 and 500°C, respectively. Some researchers have identified only Al_3Fe precipitates during recrystallization of commercial Al-Fe-Si alloy. In AA8011 alloy, Al_3Fe and Si precipitates are observed in cold rolled and annealed conditions. After heavy plastic deformation (~92%), very small particles of 0.17 μm size are distributed in the dislocations. Upon progress of annealing, the particle size increases, and both smaller and larger, spherical and plate-shaped Al_3Fe particles are inhomogeneously distributed and situated at subboundaries and subgrain interiors. Since the dislocations cannot cross the grain boundary, during annealing a large variation of dislocation density occurs from grain to grain owing to the pinning effect of precipitate particles. After completion of recrystallization, particle also inhibits the grain boundary migration along with pinning of the dislocations. Even after a long time of high temperature annealing for heavily deformed alloy, the Al_3Fe particles may pin down the dislocations, resulting in the presence of a large amount dislocations inside the grains. During the intersection of grain boundary by particles, a Zener drag force is generated to restrain the boundary migration. The coherent particle generally loses the coherency when a high angle grain boundary moves past a coherent particle. The coherent particles are twice as effective in pinning a grain boundary as incoherent particles of the same size. There are many alternative situations during particle-boundary interaction. The particle may dissolve during passage of the boundary and re-precipitate in a coherent orientation, it may reorient itself to a coherent orientation, or the boundary may cut through the particle.

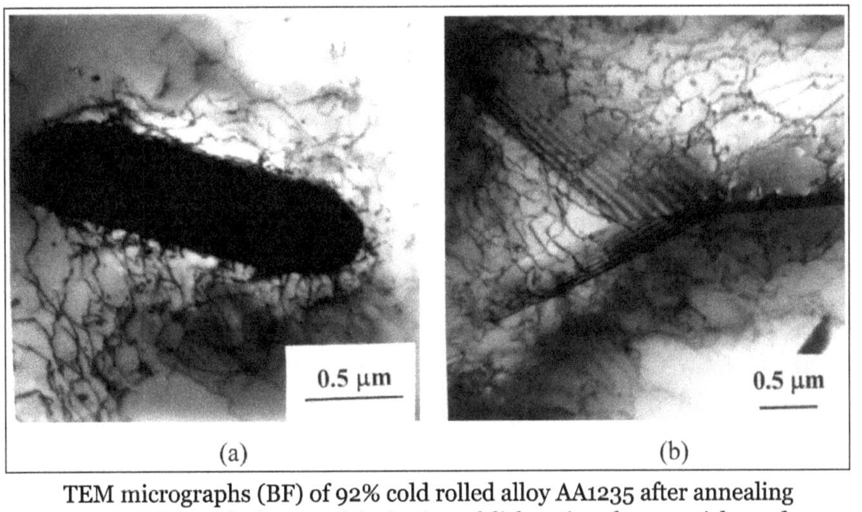

TEM micrographs (BF) of 92% cold rolled alloy AA1235 after annealing at 480 °C for 8 h showing (a) pinning of dislocations by a particle, and (b) presence of dislocations even after completion of recrystallization.

Particle Distributions in Bimodal Alloys

Many commercial alloys are bimodal alloys which contain both large (>1 μm) and small particles. Large particles act as nucleation sites for recrystallization and small particles hinder the grain boundary migration, i.e., retard the recrystallization. Therefore, recrystallization behaviour of bimodal alloys is affected by particle distribution. During deformation, the fine dispersion of small particles does not affect significantly the degree of lattice curvature generated at the large particle-matrix interface. Later, by altering the dispersion parameters of particles, recrystallization kinetics and microstructures are controlled. Chan and Humphreys have reported that in the bimodal alloy, nucleation of recrystallization takes place at the large particles, but fine particles determine the time for completion of recrystallization. They have observed a large number of small island grains and coarse irregular grains in the microstructure of bimodal alloys. With increasing coarsening of the fine precipitates, the grain shapes become more regular and the mean size decreases. Nes has taken the help of the parameter f/r (where f and r are volume fraction and radius of small particles, respectively) to describe a model to account for the grain size of bimodal alloys. In bimodal alloys, critical particle size for the growth of a nucleus is

$$d = \frac{4\gamma}{P_d - P_z} = \frac{4\gamma}{\frac{\rho G b^2}{2} - \frac{3f\gamma}{2r}}$$

where P_d is driving force (stored energy in the subgrain boundaries),

ρ is dislocation density,

G is shear modulus,

b is burgers vector,

P_z is Zener pinning force,

is grain boundary energy,

f is volume fraction of small particles,

r is radius of small particles.

As f/r of fine dispersion increases, the large particles act as the nucleation sites, but growth of the grains is slowed down at the early stage of annealing. However, the earlier developed grains either consume the smaller grains, or form island grains.

Recrystallization Texture

Annealing of heavily deformed material leads to a wide range of recrystallization textures. Different types of recrystallization textures can be produced in similar alloys each having almost identical and very strong deformation textures. In f.c.c. metals, sometimes recrystallization texture contains cube texture along with retained rolling texture and in some cases nearly random textures. It is reported that with increasing deformation, the volume fraction of the cube texture decreases preferably for low annealing temperatures and that of the rolling texture increases for all temperatures, whereas the random part of the texture strongly decreases. The possible origin of recrystallization texture has been reviewed several times. It is notable that "oriented nucleation" and "oriented growth" are two possible theories for the development of recrystallization texture. The recrystallization texture develops from the competition between cube oriented grains which nucleate at the cube bands, R-oriented grains (i.e., retained rolling texture) which stem from the grain boundaries between the former deformed grains, and randomly oriented grains due to particle stimulated nucleation (PSN).

The Effect of Particles

Widely spaced coarse particles (>1 μm) enhance nucleation and the rate of recrystallization if it is present before deformation. The finely dispersed particles normally reduce recrystallization kinetics having a greater retarding influence on nucleation than on growth. Therefore, the texture changes can be rationalized on the basis of the orientation dependent nucleation. If nucleation of all components is retarded in equal proportions, then the time available for growth of the first formed grains is increased. The recrystallization texture will therefore show increased selectivity and will become even more strongly biased towards the most favoured orientation nuclei. The effect of particle size on the formation of major components of recrystallization texture is shown in Table.

Table: Major components of the recrystallization texture for some particle containing alloys.

Material(% of reduction)	Particles		Recrystallization Texture
	Volume fraction (f)	Diameter (μm)	
Al (99.9965) (90%)	0	-	Cube
Fe-AlN (70%)	0.06	0.017	Rolling
Al-Al2O3 (30-90%)	0.4	0.1	Deformation
Cu-SiO2 (70%)	0.5	0.23	Rolling + Twins
Al-FeSi (90%)	0.5	0.2- 7	Cube, Rolling, Random
Al-Si (90%)	0.8	2	Cube, Rolling, Random
Al- Ni (80%)	10.0	1	Rolling, Random
Al- SiC (70%	20.0	10	Random

The effect of PSN on recrystallization texture is as follows:

- PSN results in a sharp recrystallization texture for low strained material or in case of particle containing single crystals. When PSN originates in several deformation zones at a single particle of a lightly deformed polycrystals, a spread of orientation is observed around the particles in the deformed matrix.
- In the case of heavily deformed material, the presence of different deformation zones of different grains or different deformation regions of the same grain results in a single grain with a wide range of orientations. As a consequence, either a weak texture or randomly oriented grains form.

The recrystallization texture is dependent on the particle size, strength and spacing. When the particles precipitate during annealing, the recrystallization texture developed is similar to the deformation texture. There is no clear reason for the retention of the rolling texture in the recrystallized alloys. It appears that a strong cube texture is rarely formed and particles may be responsible for suppressing the formation or viability of cube sites. This enables other components including the retained rolling components to dominate the texture.

Although PSN nuclei are oriented randomly in the heavily rolled polycrystal, the final texture of this alloy is not random. Therefore, it is concluded that grains from other sites with different orientation affect the final texture. From table it is observed that for the alloys (e.g., Al-Fe-Si or Al-Si) containing low volume fraction of particles, cube and rolling components are developed, whereas rolling and random components develop for high volume fraction of particles containing alloys (e.g., Al-Ni). The cube grains are larger than randomly oriented grains owing to the faster growth rate of former. It results in formation of small island grains inside the large cube grains. Some researchers have shown that the strength of the cube component decreases with an increase in the number of supercritical sized particles in the case of hot-rolled aluminium alloy AA3004. On the other hand, volume fraction of randomly oriented grains increases with an increase in the number of supercritical sized particles. Hornbogen et al. have distinguished between 'discontinuous' (conventional recrystallization) and 'continuous' recrystallization. The former cases give rise to normal recrystallization textures, such as the cube texture, while the latter cause retention of the rolling texture.

Effect of Fe on the Commercial Purity Aluminium

A typical recrystallization texture of highest purity aluminium is the cube-orientation {001}<100>. In commercial purity aluminium the iron and silicon contents have been found to be important factors in controlling the recrystallization texture. It is reported that a small amount of Fe may cause a change almost from pure cube to retained rolling texture. During recrystallization, precipitation of Al_3Fe particles inhibits the growth of the early formed cube nuclei, thereby forcing nucleation in the abundant rolling texture components. Thus, the recrystallization textures of most commercial aluminium alloys are composed of three texture types, namely the cube texture, the retained rolling texture and the random texture. Cube-oriented subgrains are known to rapidly recover either dynamically during deformation or statically during the early stage of recrystallization, which gives rise to a size advantage of cube nuclei. Growth of the dominant cube grains is lowered by solute drag and/or precipitation, which leads to the development of rolling component and reduces the strength of the cube component. In a sheet, a mixture of cube and rolling components is desired to prohibit earing during deep drawing of that sheet. It is interesting to note from the research work of Hirsch and Lücke that 95% cold-rolled alloy Al-0.007% Fe shows the retained rolling texture with a small amount of cube texture at 360°C compared to strong cube texture at 280 and 520°C. It is due to the formation of iron-rich phase at 360°C, which restricts cube grain growth. When precipitation occurs before or after completion of recrystallization, the effect on boundary migration is less drastic and a strong cube texture develops. It has been observed that the growing grains of cube component can consume the recrystallized grains of non-cube component along with the surrounding deformed material. It is reported that coarse iron rich particles inhibit the formation of cube texture in cold rolled aluminium alloys. It is evident that increasing iron content decreases the strength of the cube texture, especially when it is present in solid solution form before rolling. Even the addition of 0.1% iron to aluminium may change the recrystallization texture.

Effect of Recrystallization Texture on Microstructure and Mechanical Properties

Grain size and mechanical properties can be controlled by maintaining proper texture in recrystallized microstructure. Al_3Fe particles suppress the cube texture formation. Cube grain growth is also inhibited by "orientation pinning". Orientation pinning occurs when a recrystallized grain of a given orientation grows into deformed material of its own orientation. It results in a low angle boundary and the growth of the recrystallized grain will practically stop due to the low mobility of low angle boundaries. The case where orientation pinning is expected to be important is for the growth of grains with the same orientation as a main component in the deformation texture, like retained rolling components. Vatne et al. have also reported a case where the growth of cube grains has been most likely reduced due to orientation pinning. This may be due to an extremely high as-deformed cube fraction of 35%. Therefore, fine grains can be generated in the foil by controlling recrystallization texture during intermediate annealing treatment, which improves the mechanical properties of the foil. Blade has demonstrated that earing tendency of aluminium alloy increases with increase in cube texture component. It is shown that earing tendency becomes zero when 25% cube texture is present.

Grain Growth

Effect of Second-phase Particles and Orientation Gradient

During grain growth, boundary migration is retarded by a Zener drag effect of the second-phase particles. As the driving pressure of grain growth is extremely low, particles may have a very large influence on both the kinetics of grain growth and the resultant microstructure. Humphreys et al. have elaborately discussed the growth related formulation, i.e., driving and pinning pressures for growth, grain growth rate, limiting grain size due to particles, etc.. In the case of planar grain boundary, Zener limiting grain size (D_{Zener}) which equals to the critical radius of island grains, arises from the balance between the driving pressure for grain growth (P) and Zener pinning pressure (P_z).

$$D_{Zener} = \frac{4r}{3f}$$

The above equation is only applicable to those materials where the second-phase particles are stable during grain growth. Therefore, stability of the particles with temperature is one of the important factors for growth mechanism. The instability of the second-phase particles, i.e., precipitation after the formation of grain and subgrain structure or coarsening of particles during grain growth, affects the grain growth by a different way.

Orientation gradient has a greater effect on the grain boundary mobility. High stacking fault energy alloys (Al, Ni and α-iron etc.) readily form a cellular or subgrain structure during deformation. This structure is often not uniform and orientation gradients are usually developed here. In addition, many alloys produced by thermomechanical processing develop a preferred orientation or texture by recrystallization and grain growth. A model developed by Ferry illustrates that the grain coarsening is rapid for large orientation gradient and particle-free systems, and it is markedly reduced in a system containing a large volume fraction of fine particles, despite the presence of the orientation gradient. The latter case occurs due to retardation of the onset of recrystallization.

Competition between Normal and Abnormal Grain Growth

Grain growth may be divided into two types, normal grain growth and abnormal grain growth. When the grains grow uniformly with a narrow range of size and shapes and the grain size distribution is independent of time, then, this type of grain growth is called "normal" or "continuous" grain growth. In "abnormal" or "discontinuous" grain growth, few grains are large compared to the rest and grain size distribution is bimodal. Since this discontinuous growth of selected grains has similar kinetics to primary recrystallization and has some microstructural similarities, abnormal grain growth is sometime called secondary recrystallization. The normal grain growth theory is based on the grain boundary interfacial free energy as the driving force. If the initial grain size distribution is too wide, a fraction of large grains will grow in an abnormal manner until all the other grains have been consumed. When completed, this process has resulted in a more narrow size distribution and at longer times the steady state may be approached asymptotically. It appears convenient to define this state as a normal grain growth. Abnormal grain growth may sometime be a necessary initial stage toward find normal grain growth.

The origin of abnormal grain growth is yet to be fully understood. However, by Monte Carlo simulation it has been observed that two different conditions may initiate abnormal grain growth. Firstly, anisotropy in the grain boundary energy may lead to rapid growth of grains having boundary energies much lower than the average. This is often the case when the material exhibits a strong primary recrystallization texture. Secondly, abnormal grain growth may arise from anisotropy in the grain boundary mobility.

Hillert has deduced three conditions for the development of abnormal grain growth in a material as follows:

- Normal grain growth cannot take place due to particle pinning.
- The average grain size has a value below the limit $1/2z$(where $z = 3f/4r, f$ and r are the volume fraction and the radius of particles, respectively).
- There is at least one grain much larger than the average.

Whether these conditions are automatically fulfilled in a material where the normal grain growth has stopped growing due to the presence of second phase particles, is a question of considerable practical importance. Again, the growth of a very large grain in a material where the normal grain growth has stopped depends upon the value of the final grain size reached by the normal grain growth. The mathematical analysis, on the other hand, predicts that normal grain growth should proceed up to the limit 1/2z. This causes abnormal grain growth to be impossible for a limited time period when the normal grain growth has started to slow down but has not yet reached the limit. Therefore, initiation of abnormal grain growth is possible by a continuous decrease in the z value. This effect can be accomplished increasing the particle size r through coalescence or decreasing the volume fraction f by dissolving the second phase. Any process that leads to a slow increase in the grain size limit may initiate the development of abnormal grain growth. Most cases of abnormal grain growth, met within practice, seem to be connected with the dissolution of the second phase rather than the coalescence.

The occurrence of abnormal grain growth may be limited by "nucleation" rather than growth considerations. It is evident from a large number of publications that abonormal grain growth is likely to occur as the annealing temperature is raised and as the particle dispersion becomes unstable. If a single strong texture component is present in a fine-grained recrystallized material, then abnormal grain growth commonly occurs on further annealing at high temperatures. This is due to a low misorientation and hence low energy and mobility of grain boundaries in highly textured materials. Due to the presence of another texture component, higher energy and mobility are introduced in the boundaries to migrate preferentially by a process, which is closely related to primary recrystallization. The avoidance of abnormal grain growth at elevated temperatures is an important aspect of grain size control in steels and other alloys. A safe method of avoiding abnormal grain growth would be to form an average grain size so much larger than the particle-limited grain size. Another method is by choosing a material with a larger volume fraction of second phase particles. In other words, it can be said that abnormal grain growth is not likely to occur where most of the particles are observed to be situated at the grain boundaries.

References

- Aluminum: newworldencyclopedia.org, Retrieved 05 January, 2019
- Aluminum-or-aluminium-alloys-603707: thoughtco.com, Retrieved 14 March, 2019
- Aluminium-alloys-and-its-heat-treatment-metallurgy-26614: engineeringenotes.com, Retrieved 03 April, 2019
- Mechanical-behavior-of-precipitation-hardened-aluminum-alloys-welds, light-metal-alloys-applications: intechopen.com, Retrieved 31 May, 2019
- Aluminium-alloys-and-its-heat-treatment-metallurgy- 26614: engineeringenotes.com, Retrieved 26 June, 2019
- Understanding-the-aluminum-alloy-designation-system: esabna.com, Retrieved 14 February, 2019
- Aluminium-alloys-and-its-heat-treatment-metallurgy- 26614: engineeringenotes.com, Retrieved 15 March, 2019
- Additive-manufacturing-of-al-alloys-and-aluminium-matrix-composites-amcs, light-metal-alloys-applications: intechopen.com, Retrieved 26 July, 2019
- Recrystallization-behavior-of-commercial-purity-aluminium-alloys, light-metal-alloys-applications: intechopen.com, Retrieved 18 August, 2019

Understanding Magnesium Alloys

Magnesium is a chemical element which exists as a shiny gray solid. Its alloys are highly regarded for their combined properties of lightness and strength. It is used to make aircraft components, ladders, portable power tools and automotive wheels. This chapter discusses the diverse properties and applications of magnesium alloys in detail.

Magnesium

Magnesium is the lightest of all metal elements and is primarily used in structural alloys due to its lightweight, strength, and resistance to corrosion.

There are over 60 different minerals known to have a magnesium content of 20% or greater, making it the eighth most abundant element in the earth's crust. But when water bodies are accounted for, magnesium becomes the most abundant element on the surface of the earth. That is because of the significant magnesium content in salt water, which averages about 1290 parts per million (ppm). Yet, despite its abundance, global magnesium production is only about 757,000 tons per year.

Properties

- Atomic Symbol: Mg.
- Atomic Number: 12.
- Element Category: Alkaline metal.
- Density: 1.738 g/cm^3 (20 °C).
- Melting Point: 1202 °F (650 °C).
- Boiling Point: 1994 °F (1090 °C).
- Moh's Hardness: 2.5.

Characteristics

Magnesium's characteristics are similar to its sister metal aluminum. It not only has the lowest density of all metal elements, making it the lightest, but it is also very strong, highly resistant to corrosion and easily machinable.

Elemental magnesium is a fairly strong, lightweight metal (two thirds the density of aluminum). It slightly tarnishes when exposed to air, although unlike the alkaline metals, storage in an oxygen free environment is unnecessary because magnesium is protected by a thin layer of oxide that is fairly impermeable and hard to remove.

Like, calcium, magnesium reacts with water at room temperature, although it reacts much more slowly than calcium. When it is submerged in water, hydrogen bubbles begin to form almost unnoticeably on the metal's surface, but the powdered metal reacts much more rapidly. At higher temperatures, the reaction will occur faster.

Magnesium is a highly flammable metal that is easy to ignite when powdered or shaved into thin strips. It is, however, difficult to ignite in mass or bulk. Once ignited, it is difficult to extinguish, being able to burn in both nitrogen (forming magnesium nitride), and carbon dioxide (forming magnesium oxide and carbon). It burns at a temperature of approximately 2,500 K (2200 °C, 4000 °F). The autoignition temperature of magnesium is approximately 744 K (473 °C, 883 °F).

When it burns in air, magnesium produces a brilliant white light. This property was applied in the early days of photography, when magnesium powder (flash powder) was used as a source of illumination. Later, magnesium ribbon was used in electrically ignited flash bulbs.

Physical properties	
Phase	solid
Density (near r.t.)	1.738 g/cm³
Liquid density at m.p.	1.584 g/cm³
Melting point	923 K (650 °C, 1202 °F)
Boiling point	1363 K (1090 °C, 1994 °F)
Heat of fusion	8.48 kJ/mol
Heat of vaporization	128 kJ/mol
Heat capacity	(25 °C) 24.869 J/(mol·K)

Vapor pressure						
P/Pa	1	10	100	1 k	10 k	100 k
at *T*/K	701	773	861	971	1132	1361

When glowing white at high temperatures, magnesium has many chemical properties that it does not possess at lower temperatures. It also becomes more toxic, but the high temperature alone is extremely dangerous.

Magnesium compounds are typically white crystals. Most are soluble in water, providing the sour-tasting magnesium ion Mg^{2+}. Small amounts of dissolved magnesium ion contributes to the tartness and taste of natural waters. In large amounts, magnesium ion is an ionic laxative. The so-called "milk of magnesia" is a water suspension of one of the few insoluble magnesium compounds, magnesium hydroxide, and it is named after the appearance of the undissolved particles. Milk of magnesia is a mild base.

Isotopes

Three isotopes of magnesium are found in nature: ^{24}Mg, ^{25}Mg, and ^{26}Mg, at abundances of approximately 79%, 10%, and 11%, respectively. All three isotopes are stable.

^{26}Mg has found application in isotopic geology, similar to that of aluminum. ^{26}Mg is a radiogenic daughter product of ^{26}Al, which has a half-life of 717,000 years. Large enrichments of stable ^{26}Mg have been observed in the Ca-Al-rich inclusions of some carbonaceous chondrite meteorites. The anomalous abundance of ^{26}Mg is attributed to the decay of its parent ^{26}Al in the inclusions. Therefore, the meteorite must have formed in the solar nebula before the ^{26}Al had decayed.

Atomic properties	
Crystal structure	hexagonal
Oxidation states	2 (strongly basic oxide)
Electronegativity	1.31 (Pauling scale)
Ionization energies (more)	1st: 737.7 kJ/mol
	2nd: 1450.7 kJ/mol
	3rd: 7732.7 kJ/mol
Atomic radius	150 pm
Atomic radius (calc.)	145 pm
Covalent radius	130 pm
Van der Waals radius	173 pm

Applications

Uses of the Metal and its Alloys

Magnesium is the third most commonly used structural metal, following steel and aluminum. In its purest form, magnesium can be compared to aluminum. Like aluminum, it is strong and light. Its principal use is as an alloying additive to aluminum—these aluminum-magnesium alloys, sometimes called *magnalium* or *magnelium*, are used mainly for beverage cans.

Magnesium alloy is also used in several high-volume parts manufacturing applications, including automotive and truck components. Specialty, high-grade car wheels of magnesium alloy are called "mag wheels." In 1957, a Corvette SS, designed for racing, was constructed with completely magnesium body panels. Volkswagen has used magnesium in its engine components for many years. For a long time, Porsche used magnesium alloy for its engine blocks, due to the weight advantage.

Recently, there appears to be renewed interest in magnesium engine blocks, as featured in the 2006 BMW 325i and 330i models. The award-winning BMW engine uses an aluminum alloy insert for the cylinder walls and cooling jackets surrounded by a high-temperature magnesium alloy AJ62A. The application of magnesium AE44 alloy in the 2006 Corvette Z06 engine cradle has advanced the technology of designing robust automotive parts in magnesium. Both these alloys are recent developments in high-temperature, low-creep magnesium alloys. New alloy development and lower costs, which are becoming competitive with aluminum, will increase the number of automotive applications.

Magnesium is also used in electronic devices. Given its low weight and good mechanical and electrical properties, it is widely used in the manufacture of mobile phones, laptops, cameras, and so forth.

There are several other uses of magnesium, some of which are listed below:

- It is a reducing agent for the production of pure uranium and other metals from their salts. It is also used for the removal of sulfur from iron and steel.
- The printing industry uses it for photoengraved plates.
- As an alloying agent, it improves the mechanical, fabrication, and welding characteristics of aluminum.
- Combined in alloys, it is essential for airplane and missile construction.
- It is an additive agent for conventional propellants and used in producing nodular graphite in cast iron.
- The extremely high temperature at which magnesium burns makes it a handy tool for starting emergency fires during outdoor recreation.
- As burning magnesium produces a brilliant white light, it is used in flashlight photography, flares, pyrotechnics, sparklers, and incendiary bombs.

Uses of Magnesium Compounds

- Magnesium compounds, primarily magnesium oxide, are used mainly as refractory materials in furnace linings for producing iron, steel, nonferrous metals, glass, and cement.
- The hydroxide of magnesium, or milk of magnesia, is used as a laxative and antacid; its chloride and citrate are used as oral magnesium supplements; and its sulfate (Epsom salts) is used for various purposes in medicine.
- Magnesium carbonate ($MgCO_3$) is used in making bricks and in flooring, fireproofing, and fire-extinguishing compositions. Its powder is used by gymnasts and weightlifters to improve their grip on objects such as the lifting bar.
- Magnesium stearate is a slightly flammable white powder with lubricative properties. The pharmaceutical industry uses it in the manufacture of tablets, to prevent the tablets from sticking to the equipment during the tablet compression process.
- Given that magnesium (in the ionic form) is necessary for living organisms, magnesium salts are additives in foods, fertilizers, and tissue culture media.

Precautions

Pure magnesium metal and its alloys are highly flammable when in the molten state, powdered form, or ribbon form. Burning or molten magnesium metal reacts violently with water. Magnesium powder is an explosive hazard. One should wear safety glasses while working with magnesium. The bright white light (including ultraviolet) produced by burning magnesium can damage the eyes. Water should not be used to extinguish magnesium fires, because it can actually feed the fire, according to the following reaction:

$$Mg_{(s)} + 2\,H_2O_{(v)} \rightarrow Mg(OH)_{2\,(aq)} + H_{2\,(g)}$$

or in words:

$$\text{Magnesium}_{(solid)} + \text{steam} \rightarrow \text{Magnesium hydroxide}_{(aqueous)} + \text{Hydrogen}_{(gas)}$$

Carbon dioxide fire extinguishers should not be used either, because magnesium can burn in carbon dioxide (forming magnesium oxide, MgO, and carbon). A dry chemical fire extinguisher (Class D) should be used if available, or else the fire should be covered with sand or magnesium foundry flux. An easy way to put out small metal fires is to place a polyethene bag filled with dry sand on top of the fire. The heat of the fire will melt the bag and the sand will flow out onto the fire.

Magnesium Alloys

Pure Mg is produced by electrolysis of $MgC1_2$, or by silico-thermic reduction of MgO, and generally show a purity exceeding 99.9 wt%. In its pure form the metal is soft and can only be hardened by deformation. For structural applications a variety of alloys have been developed. The alloying elements are added both to secure adequate processability of the metal, and to obtain the performance required by the various applications.

Alloying Elements

A basic requirement for an alloying element to be used in conventionally liquid metallurgy processed alloys, is that it shows a sufficiently high solubility in liquid magnesium. The periodic table, fig, shows elements which possess extensive, limited or negligible solubility in liquid Mg. The solubility will of course be changed in ternary and higher order systems, however, the binary systems indicate the general behaviour of the elements. The solid solubility is another important property of an alloying element, which strongly influences its effect on alloy properties. In practice the major alloying elements of interest are limited to (in order of increasing atomic number): Li, Al, Si, Ca, Mn, Cu, Zn, Sr, Y, Zr, Ag and Rare Earth (RE) elements. Some minor alloying elements are used, even if both the liquid and solid solubility are negligible. These elements include C, Be and Ti.

The overall properties of a fabricated Mg-alloy component depends upon its microstructure, which is determined through a complex interaction between the alloy chemistry and the processing parameters. Salient features include type, amount and distribution of elements in solid solution, age hardening precipitates, dispersoids, intermetallic constituents, dendrite arm spacing, grain size and shape, metallic and nonmetallic impurity elements.

In the following paragraphs some basic aspects of commercial Mg-alloy processing are discussed, followed by examples of recent developments.

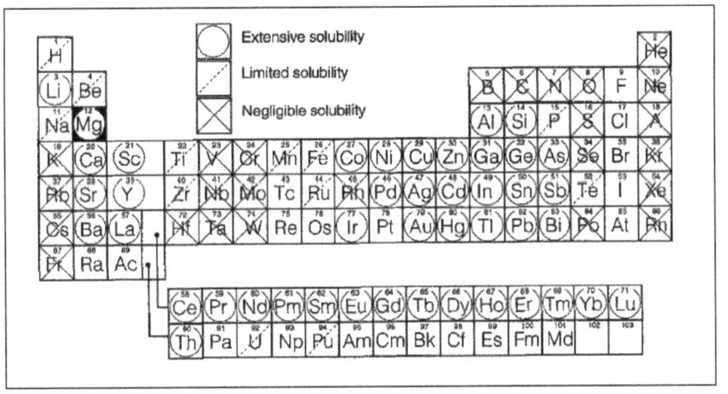

Periodic table indicating elements with extensive solubility, limited solubility and negligible solubility in liquid magnesium.

Control of Heavy Element Impurities

The impurities Fe, Ni and Cu strongly influence the corrosion properties of Mg-alloys. No simple ways are known to remove Ni and Cu since both elements exhibit extensive liquid solubility. Hence these elements must be limited by proper control of raw materials, and any Ni and Cu contamination must be avoided during processing. Current high purity standards limit the content of these elements to 10 and 300 ppm respectively. The Fe-level can be controlled by liquid metal treatment aimed at precipitating the Fe by adding elements which reduce its liquid solubility. For this purpose Mn is commonly used. It is generally agreed that the corrosion mechanism is microgalvanic in nature caused by rnicrocells, where Fe-rich intermetallics constitute the cathode. The galvanic current, measured during potentiostatic polarization of isolated intermetallic particles, provides a measure of the impact on corrosion properties of the various phases, fig. It was observed that intermetallic compounds, containing more than a few percent Fe are detrimental, because they function as efficien cathodes. On the other hand, binary AI-Mn phases with a low Al/Mn ratio may also exhibit a relatively high cathodic current output. It can be concluded that the best corrosion properties are obtained by keeping the Mn-additions at a level which is necessary to reduce the Fe-content (<50 ppm), and at the same time minimize formation of Mn-rich particles. Other elements which precipitates Fe include Zr, combinations of Zr and Si, Ti and RE-elements.

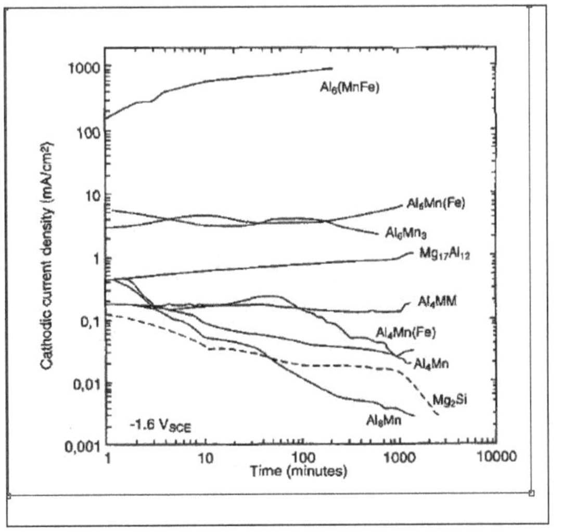

Cathodic current density of intermetallic compounds vs. time.
Applied potential of - 1.6 V,,, in deaerated 5 % NaCl saturated with $Mg(OH)_2$.

Control of Melt Reactivity

Protective gases (mixtures of SF_6, CO_2, Ar and air) as well as salt containing fluxes are used to prevent oxidation of the liquid alloys. Due to the adverse effect on working environment, disposal problems, as well as a possible deterioration of metal cleanliness due to salt inclusions, gas protection is now the principal melt protection method. The oxidation rate is also influenced by alloying elements. Al-additions reduce the tendency to melt oxidation, and Be is strongly reducing oxidation at a few ppm level (5- 15 ppm). Other elements such as RE, Li, Ca and Sr increase the oxidation rate, and special precautions must be taken (gas protection).

Control of Grain Size

Control of the grain structure is a key factor during solidification processing of metals and alloys. The spatial distribution of alloying elements and phases present, as well as the hot cracking tendency and distribution of porosity are strongly influenced by the grain structure. A variety of grain refining methods are employed for casting of Mg. These include constitutional grain refining, rapid cooling, Zr-additions and carbon inoculation. For alloys containing Al and Mn as alloying elements, a common grain refining method is to add hexachlorethane, C_2Cl_6. However, due to environmental concerns regarding emission of chlorinated hydrocarbons, this method is questionable. A new grain refining additive using the principle of carbon inoculation has been developed. The grain refining effect of this additive is illustrated in fig. Although the effect is most pronounced at higher melt temperatures, satisfactory results are obtained also at lower melt temperatures after considerable holding times.

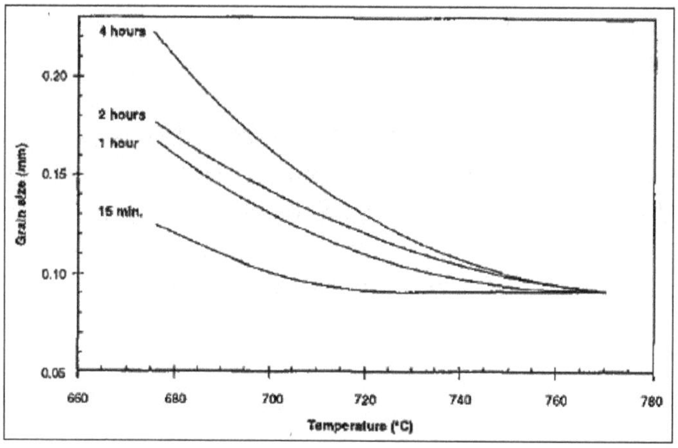

Grain size of carbon-inoculated AZ91 D vs. melt temperature at various holding times.

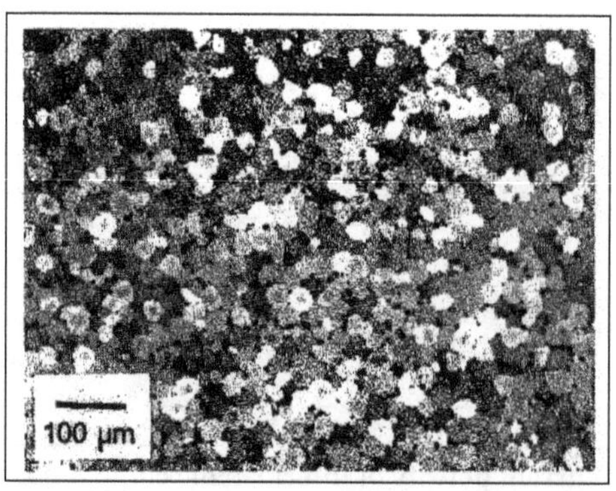

Grain structure of Zr treated alloy ZE41.

For alloys which do not contain Al or Mn, Zr-additions are commonly used for grain refining. Zr-inoculation is extremely efficient, and the resulting grain structure is completely nondendritical, figure. The grain size is determined by the cooling rate in a similar manner as the secondary dendrite arm spacing.

Commercial Alloy Systems

Mg-alloys, including cast and wrought alloys can be broadly divided into two groups: Magnesium-Aluminium alloys and Zirconium containing alloys.

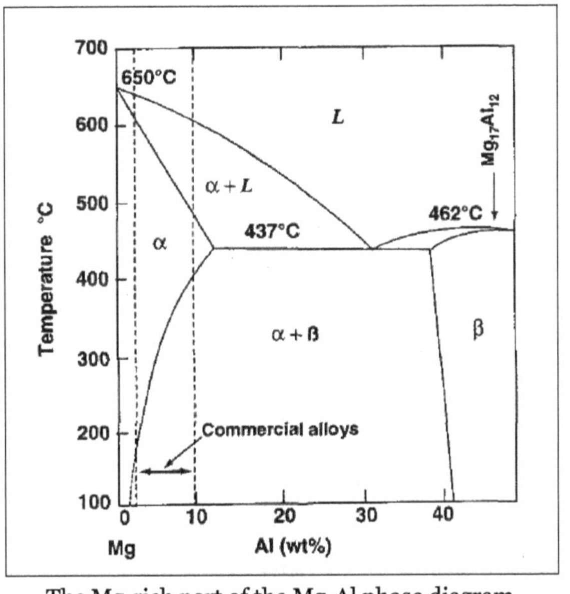

The Mg-rich part of the Mg-Al phase diagram.

Table: MG-based casting alloys. Nominal chemical composition (wt%).

Designation	Al	Zn	Mn	Si	Zr	Y	Ag	Other
AM20	2.0	<0.1	0.4	<0.1				
AM50	5.0	<0.1	0.2	<0.1				
AM60	6.0	<0.2	0.25	<0.1				
AS21	2.2	<0.1	0.2	1.0				
AS41	4.5	<0.1	0.2	1.0				
AE42	4.0	<0.2	0.25	<0.1				2.0-3.0 RE(Ce-rich)
AZ63	6.0	3.0	0.15	<0.3				
AZ91	9.0	0.7	0.15	<0.1				
EZ33		2.7			0.6			3.3 RE(Ce-rich)
ZE41		4.2			0.7			1.2 RE(Ce-rich)
EK31					0.7			3.7 RE(Nd-rich)
EQ21					0.5		1.5	2.1 RE(Nd-rich)
QE22					0.7		2.5	2.1 RE(Nd-rich)
WE54					0.5	5.3		3.5 RE(Nd-rich)
WE43					0.5	4.0		3.0 RE(Nd-rich)

Table: Mg-based wrought alloys. Nominal chemical composition (wt%).

Designation	Al	Zn	Mn	Si	Zr	Y	Ag	Other
M1 A			1.5	<0.1				
AZ31 B	3.0	1.0	0.20	<0.1				
AZ61 A	6.5	1.0	0.15	<0.1				
AZ80 A	8.5	0.5	0.12	<0.1				
ZK30 A	3.0				0.6			
ZK60 A	5.5				0.4			
WE43					0.5	4.0		3.0 RE(Nd-rich)

Magnesium-Aluminium Alloys

The majority of Mg-alloys contains Al as the main alloying addition. A section of the binary Mg-Al phase diagram is shown in fig. For practical purposes, the Al-content is limited to about 10 wt% due to the formation of a brittle eutectic containing about 72 wt% of the P-Mg,,Al,, phase. This means that Mg-Al casting alloys have compositions, which normally are considered as typical for wrought alloys. Unlike AlSi casting alloys, the Mg-Al based casting alloys can be heat treated into a single phase material. The consequence is that Mg-Al alloys are somewhat less castable than Al-Si alloys, however, a superior ductility can be obtained. In practical casting processes, solidification occurs at relatively high cooling rates, and according to the Scheil equation, a relatively large fraction of eutectic is formed, even at lower Al-levels. This secures the necessary castability of the alloys. Mn and Zn are corn monly added at levels 0.2-0.3 and 0.5-2 wt%, res pectively, to improve corrosion resistance and strength. A typical microstructure of as cast (permanent mould) AZ91D is shown in fig., where β -Mg_{17},Al_{12} AlMn(Fe) intemetallics, and discontinuously precipitated β-Mg_{17},Al_{12} can be seen. Following a solution and amficial ageing heat treatment, platelets of semicoherent β-$Mg_{17}Al_{12}$ are formed on the planes of the Mg-matrix, resulting in a moderate hardening effect.

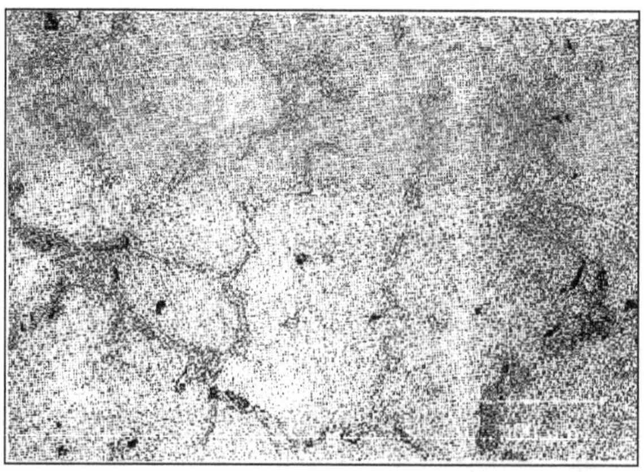

Microstructure of permanent mould cast AZ91 D.

Figure illustrates the remarkable ductilities which can be obtained in high pressure die cast specimens, by reducing the level of Al and hence the volume fraction of eutectic constituents.

For applications exposed to working temperatures exceeding about 120 °C, the Mg-Al alloys may show inadequate creep properties. A reduction in the creep rate is obtained by addition of Si, which has a negligible solid solubility, causing formation of a fine dispersion of Mg_2Si precipitates, figure. This led to the introduction of the creep resistant alloys AS41 and AS21.

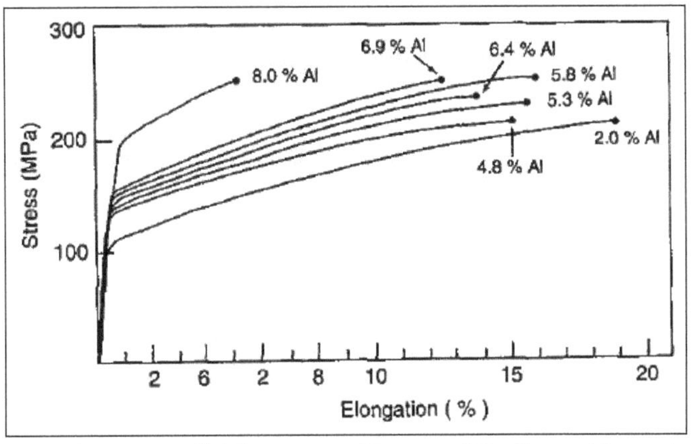

Nominal stress-strain curves of high pressure die casts AM-alloys with varying Al-content.

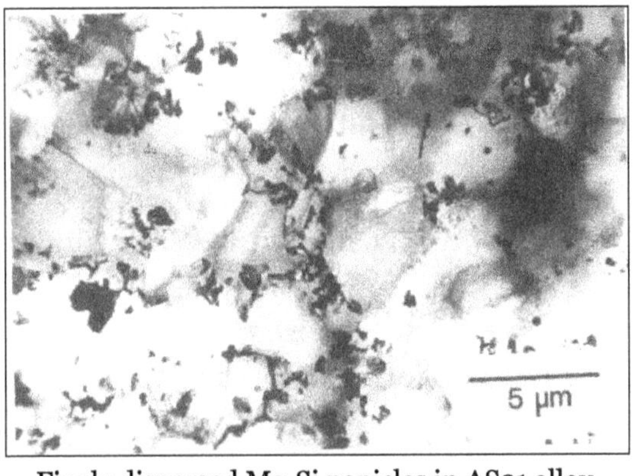

Finely dispersed Mg_2Si panicles in AS21 alloy.

More recently, alloys containing RE-elements, e.g. Ce, La and Nd ('mischmetal') have been introduced. In these alloys a dispersion of Al_4RE panicles is formed, while simultaneously reducing the amount of the low-melting eutectic a(Mg) + $\beta(Mg_{17}Al_{12})$. At elevated temperatures, these eutectic regions soften considerably due to the high homologous temperatures. The alloy AE42 is currently undergoing extensive testing for applications exposed to elevated temperatures due to the favourable creep properties, fig.

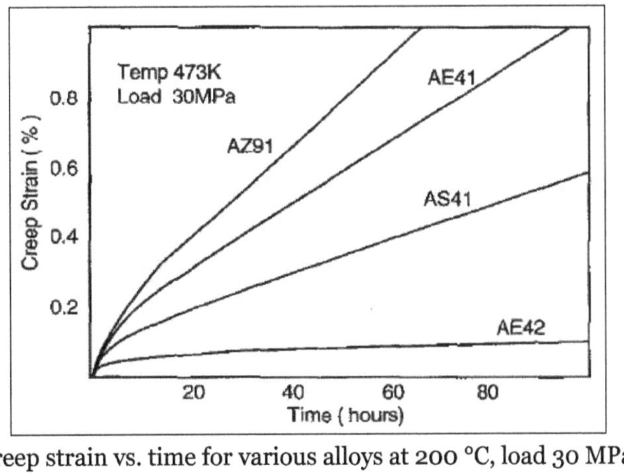

Creep strain vs. time for various alloys at 200 °C, load 30 MPa.

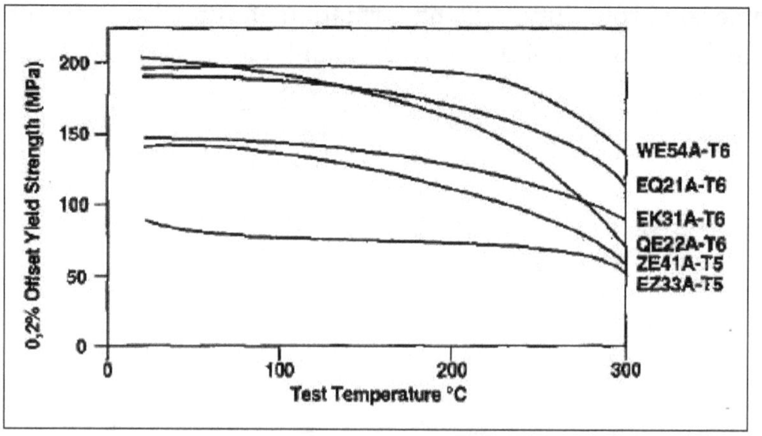

Temperature dependence of the 0.2 % offset yield strength for various Zr grain refined alloys.

Zirconium Grain Refined Alloys

Due to the extremely efficient grain refining effect of Zr, a series of alloys, especially suited for sand casting, has been developed. Unfortunately, the presence of elements like Al and Mn suppresses the liquid solubility of Zr, precluding the exploitation of this grain refining mechanism in the common Mg-Al alloys.

Elements compatible with Zr include Zn, RE, Ag and Y. While binary Mg-Zn alloys show inferior mechanical properties and castability to Mg-Al alloys, effective grain refinement by Zr, combined with additions of RE-elements to reduce microporosity, led to the introduction of alloys such as EZ33 and ZE41. The ZE63 alloy is a high strength variant, for which solution heat treatment in hydrogen atmosphere is employed to remove some of the embrittling MgZnRE grain boundary phases by formation of RE-hydrides. At the same time Zn becomes available for age hardening. An age hardening system containing Ag, Nd-rich mischmetal (a mixture of rare earth elements) and Zr, shows high mechanical strength and good creep properties at temperatures approaching 250 °C (EQ21 and QE22). The most recent alloy system, which has been evaluated, is the WE series, containing Y and Nd-rich mischmetal in addition to Zr. The WE54 alloy shows considerable strength up to 300 °C, fig. Due to a certain embrittling effect caused by long term ageing at elevated temperatures, a slight reduction in alloying content has been suggested, and WE43 has now been introduced. These alloys are used in the fully aged condition.

Mechanical Properties of Magnesium Alloys

The mechanical properties of magnesium alloys are determined in accordance with ASTM procedures where applicable. Properties of specification and commercial grades will meet the minimum specified values established by the principal producers. Minimum properties represent values of specification and commercial grade alloys when such values have been established. In cases where they have not been established, the minima are given only as suggested values for design purposes.

As is the case with other nonferrous metals, magnesium exhibits no definite yield point. In accordance with ASTM procedure, the yield strength of magnesium alloys is defined as the stress at which the stress-strain curve deviates 0.2 percent from the modulus line (the so-called 0.2 percent offset point).

The compressive yield strength of magnesium alloys is defined as the stress necessary to produce 0.2 per cent permanent set. In the case of cast magnesium alloys the compressive yield strength is substantially the same as the tensile yield strength. However, for most wrought magnesium alloys the compressive yield strength is 60 to 70 percent of the tensile yield strength. Compressive properties are determined on specimens having a slenderness ratio (1/g) of 12.

Casting Alloys: Tensile and other mechanical properties of the cast alloys are determined on separately poured test bars conforming to standard ASTM procedure. Specifications may require that specimens cut from sections of industrial castings have tensile ultimate and yield strengths that are at least 75 percent of those determined on separately cast bars and elongation at least 25 percent. For small grain size in both bars and castings, the strength values of sections up to 1/2 in. thick may be about 90 percent of the minimum test-bar values.

Wrought Alloys: As maybe noted, some alloys are produced in several wrought forms while others find application for one kind of manufacture only, for example, extrusions or forgings.

The mechanical properties of wrought magnesium alloys are determined on specimens cut from the actual manufactures - extrusions, forgings, or rolled products. Specimens are usually cut in a direction parallel to the direction of working or predominant flow of metal. They may also be cut transversely or at 45° to show the directional properties.

The properties given for sheet generally apply to material in the range of thickness from 0.020 to 0.250 in. Flat rolled material which is 0.251 in. and upwards in thickness is defined as plate. The mechanical properties of strip cut from sheet are the same as those of the sheet, and the properties of coiled strip approximate those of sheet. Properties for extrusions are specified separately for solid shapes (including bars and rods), for hollow shapes, and for tubing.

In the case of forgings, sections worked from extruded stock 3 in. in diameter or less usually exhibit tensile and compressive yield strengths greater than the minimum specification values. The mechanical properties of forgings made from thicker stock which is subjected to substantial movement of metal in the processing may also exceed the specified minima, but ordinarily by lesser margins than do small forgings. However, thick, heavy forgings made from cast billets or extruded stock 6 to 8 in. thick may have properties considerably below the average because of the difficulty of effecting enough mechanical work during either extrusion or forging. Actually, the properties obtainable in a particular forging may be varied appreciably by variations in the production practice.

Directional Properties: In the case of wrought magnesium alloys, the mechanical properties may vary appreciably depending upon the degree of working and the direction of metal flow. The lowest tensile properties usually occur in the "across-grain" direction (that is, perpendicular to the direction of working) and in sections that have been worked the least. Thus, relatively poor properties may appear in thick sections and in those subjected to the least amount of flow in the processing. It is to be noted, however, that the longitudinal properties of sheet are normally a little lower than the transverse properties.

For specification purposes, the tensile properties of sheet and plate in magnesium-base materials are ordinarily determined on specimens cut parallel to the direction of rolling, that is, longitudinal specimens. As indicated just above, properties in this direction (with but few exceptions) are slightly lower than those of transverse specimens. Likewise, in the case of extrusions, the tensile properties are normally determined on longitudinal specimens. Generally, they have properties higher than those of transverse specimens. In forgings, the tensile properties are determined on specimens taken in the direction of maximum metal flow. These properties are usually higher than properties determined in other directions of the forging.

In wrought magnesium alloys, it has been found that high yield strength in tension is generally accompanied by low yield strength in compression, and vice versa. For example, an extruded shape has a high tensile yield strength but a low compressive yield strength in longitudinal specimens; in transverse specimens this relation may often be the opposite. A similar condition has been found to exist in forgings.

Yield Strength Ratios: The ratio of compressive yield strength to tensile yield strength in magnesium materials may be influenced markedly by various factors including alloy composition, heat treatment, condition (whether cast or wrought), and directional effects. It has been pointed out that for cast magnesium alloys the compressive yield strength is about the same as the tensile yield strength. For wrought alloys, however, the former may be considerably lower.

The yield strengths of wrought magnesium alloys normally vary as a function of the direction of metal flow. In longitudinal specimens the yield strength in compression is generally about 60 to 70 percent of the yield strength in tension.

Modulus of Elasticity: For magnesium and its commercial alloys the modulus of elasticity in tension and compression is about 45 GPa. The modulus of elasticity in shear, or modulus of rigidity, is about 16 GPa. Poisson`s ratio is 0.35. In the case of Ml type magnesium alloys the modulus of elasticity has been reported as somewhat lower than 45 GPa, the value of 38 GPa having been given. The modulus of elasticity decreases with rising temperature.

Cyclic cold working to stresses exceeding the yield strength in both tension and compression reduces the modulus of elasticity of magnesium alloys. Values approximating 27.5 GPa have been observed. Such cyclic cold working, probably in the form of roll straightening, is believed to be responsible for the lower modulus values of magnesium sheet and extruded shapes which have occasionally been reported.

Endurance Limit: Considerable data are available on the fatigue strength of magnesium alloys. As is the case with metals in general, these materials are sensitive to notches and other stress-raisers which have the effect of reducing substantially their endurance limits. Unsoundness (micro shrinkage), which greatly lowers the tensile strength of cast magnesium alloys, reduces their endurance limit to similar extent.

Impact Resistance: The standard Charpy and Izod impact tests as carried out on notched bars serve to measure the relative notch sensitivity of magnesium alloys under impact loading. Similar impact tests on unnotched bars give much higher values and may be adapted to indicate resilience or toughness. Resilience is the capacity of a material to absorb shock within the elastic range. It is inversely proportional to the modulus of elasticity. Since magnesium-base materials have relatively low modulus of elasticity, they have high unit resilience. Toughness is determined by the strength and ductility. It is measured by the area under the stress-strain curve. An approximation for toughness value may be made by multiplying the tensile strength by the elongation.

Heat Treating of Magnesium Alloys

Magnesium alloys usually are heat treated either to improve mechanical properties or as means of conditioning for specific fabricating operations. The type of heat treatment selected depends on alloy composition and form (cast or wrought), and on anticipated service conditions.

Solution heat treatment improves strength and results in maximum toughness and shock resistance. Precipitation heat treatment subsequent to solution treatment gives maximum hardness and yield strength, but with some sacrifice of toughness. As applied to castings, artificial aging without prior solution treatment or annealing is a stress-relieving treatment that also somewhat increases tensile properties. Annealing of wrought products lowers tensile properties considerably and increases ductility, thereby facilitating some types of fabrication. Modifications of these basic treatments have been developed for specific alloys, to obtain the most desirable combinations of properties.

The mechanical properties of most magnesium casting alloys can be improved by heat treatment. Casting alloys can be grouped into six general classes of commercial importance on the basis of composition, as follows:

- Magnesium-aluminum-manganese.
- Magnesium-aluminum-zinc.
- Magnesium-zinc-zirconium.
- Magnesium-rare earth metal-zinc-zirconium.
- Magnesium-rare earth metal-silver-zirconium, with or without thorium.
- Magnesium-thorium-zirconium, with or without zinc.

In most wrought alloys, maximum mechanical properties are developed through strain hardening, and these alloys generally are either used without subsequent heat treatment or merely aged to a T5 temper. Occasionally, however, solution treatment, or a combination of solution treatment with strain hardening and artificial aging, will substantially improve mechanical properties.

Wrought alloys that can be strengthened by heat treatment are grouped into four general classes according to composition:

- Magnesium-aluminum-zinc.
- Magnesium-thorium-zirconium.
- Magnesium-thorium-manganese.
- Magnesium-zinc-zirconium.

Types of Heat Treatment

Annealing: Wrought magnesium alloys in various conditions of strain hardening or temper can be annealed by being heated at 290 to 455 °C (550 to 850 °F), depending on alloy, for one or more hours. This procedure usually will provide a product with the maximum anneal that is practical.

Because most forming operations on magnesium are done at elevated temperature, the need for fully annealed wrought material is less than with many other metals.

Stress Relieving of Wrought Alloys: Stress relieving is used to remove or reduce residual stresses induced in wrought magnesium products by cold and hot working, shaping and forming, straightening, and welding.

When extrusions are welded to hard rolled sheet, the lower stress-relieving temperature and the longer time should be used to minimize distortion-for example, 150 °C (300 °F) for 60 min, rather than 260 °C (500 °F) for 15 min.

Stress Relieving of Castings: The precision machining of castings to close dimensional limits, the necessity of avoiding warp age and distortion, and the desirability of preventing stress-corrosion cracking in welded magnesium-aluminum casting alloys make it mandatory that cast components be substantially free from residual stresses. Although magnesium castings do not normally contain high residual stresses, the low modulus of elasticity of magnesium alloys means that comparatively low stresses can produce appreciable elastic strains.

Residual stresses may arise from contraction due to mold restraint during solidification, from no uniform cooling after heat treatment, or from quenching. Machining operations also can result in residual stress and require intermediate stress relieving prior to final machining.

Solution Treating and Aging: In solution treating of magnesium-aluminum-zinc alloys, parts should be loaded into the furnace at approximately 260 °C (500 °F) and then raised to the appropriate solution-treating temperature slowly, to avoid fusion of eutectic compounds and resultant formation of voids. The time required to bring the load from 260 °C to the solution-treating temperature is determined by the size of the load and by the composition, size, weight and section thickness of the parts, but 2 h is a typical time.

During aging, magnesium alloy parts should be loaded into the furnace at the treatment temperature, held for the appropriate period of time, and then cooled in still air. There is a choice of artificial aging treatments for some alloys; results are closely similar for the alternative treatments given.

Reheat Treating: Under normal circumstances, when mechanical properties are within expected ranges and the prescribed best treatment has been carried out, reheat treating is seldom necessary. However, if the microstructures of heat treated castings indicate too high a compound rating or if the castings have been aged excessively by slow cooling after solution treating, reheat treating is called for. Most magnesium alloys can be reheating treated with little danger of germination.

Effects of Major Variables

Casting size and section thickness, relation of casting size to volume capacity of the furnace, and arrangement of castings in the furnace are mechanical considerations that can affect heat treating schedules for all metals.

Section Size and Heating Time: There is no general rule for estimating time of heating per unit of thickness for magnesium alloys. However, because of the high thermal conductivity of these alloys, combined with their low specific heat per unit volume, parts reach soaking temperature quite rapidly. The usual procedure is to load the furnace and then begin the soaking period when the loaded furnace reaches the desired temperature.

In the heat treating of magnesium alloy castings with thick sections a good rule is to double the time at the solution treating temperature. For example, the usual solution treatment for AZ63A castings is 12 h at about 385 °C (725 °F), whereas 25 h at about 385 °C is suggested for castings with section thickness greater than 50 mm.

Similarly, the suggested solution-treating schedule for preventing excessive grain growth in AZ92A castings is 6 h at about 405 °C, 2 h at about 350 °C and 10 h at about 405 °C; but for castings with sections more than 50 mm thick, it is recommended that the last soak at 405 °C be extended from 10 h to 19 h. The best way to determine whether or not additional solution treating time is required is to cut a section through the thickest portion of a scrap casting and examine the center of the section microscopically: If heat treatment is complete, this examination will reveal a low compound rating.

Protective Atmospheres: Although magnesium alloys can be best treated in air, protective atmospheres are almost always used for solution treating. Government specification for heat treating of magnesium castings requires a protective atmosphere for solution treating above 400 °C (750 °F). Protective atmospheres serve the dual purpose of preventing surface oxidation (which, if severe, can decrease strength) and of preventing active burning should the furnace exceed proper temperature.

The two gases normally used are sulfur dioxide and carbon dioxide. Inert gases also may be used; however, in most instances, these gases are not practical because of higher cost. Sulfur dioxide is available bottled, while carbon dioxide may be obtained either bottled or as the product of recirculated combustion gases from a gas-fired furnace. A concentration of 0.7% (0.5% min) sulfur dioxide will prevent active burning to a temperature of 565 °C (1050 °F), provided that melting of the alloy has not occurred. Carbon dioxide in a concentration of 3% will prevent active burning to 510 °C (950 °F), and a carbon dioxide concentration of 5% will provide protection to about 540 °C (1000 °F).

Equipment and Processing

In solution treating and artificial aging of magnesium alloys, it is standard practice to use an electrically heated or gas-fired furnace equipped with a high-velocity fan or comparable means for circulating the atmosphere and promoting uniformity of temperature. However, because the atmosphere for solution treating sometimes contains sulfur dioxide, only furnaces that are gastight and that provide an inlet for introducing protective atmosphere are suitable.

Quenching Media: Magnesium alloy products normally are quenched in air following solution treatment. Still air usually is sufficient; forced-air cooling is recommended for dense loads or for parts that have very thick sections.

Dimensional Stability

In normal service up to approximately 95 °C (200 °F), all magnesium casting alloys exhibit good dimensional stability and can be considered free from additional dimensional changes.

Some cast magnesium-aluminum-manganese and magnesium-aluminum-zinc alloys in certain tempers exhibit slight permanent growth after relatively long exposure to temperatures exceeding 95 °C. This growth, although slight, can give rise to problems.

In contrast to the growth characteristics of the magnesium-aluminum-zinc alloys are those of the magnesium alloys containing thorium, rare earth metals and zirconium as major alloying elements. These alloys normally are used in the T5 or T6 temper, and they shrink, rather than grow, on exposure to elevated temperatures.

Materials and Processing Designs of Magnesium Alloys

Characteristics of Mg_2Si Intermetallics

Mg_2Si bulky materials, which are formed in solid-state by consolidating Mg-33.33 mol%Si powder mixture via hot pressing process, have good mechanical properties, such as 350~700 Hv micro-hardness, high elastic modulus of 120 GPa and a 1358 K high melting point. They are much better than those of the conventional magnesium alloys. It suggests that the mechanical properties of the magnesium alloys are possible to be improved by the uniform distribution of fine Mg_2Si particles in the matrix.

Current density of the anode corresponds to the corrosion reaction speed of the materials. That is, the lower current density means more noble corrosion resistance. Compared to the conventional AZ31 magnesium alloy, the current density of Mg_2Si bulky materials is reduced to 1/40~1/60. The increase of the potential balanced between cathode and anode reaction also means that Mg_2Si is superior potential to AZ31 alloys and improves the corrosion resistance in contacting with other metals. Furthermore, the salt spray test based on JIS Z 2731 (308 K temperature and 5% concentration) reveals that the corrosion starts after 1 h in the case of AZ31 alloys. The corrosion phenomenon of the conventional magnesium alloys starts in about 1 h in salt spray test.

Mg_2Si bulky material, however, has no damage after 2,000 h endurance evaluation. It means that the corrosion resistance of Mg_2Si is much better than the conventional stainless steels.

Materials Design for High-performance Magnesium Alloys

It is possible to supply some characteristics, which are required as structural components, to magnesium alloys by using Mg_2Si intermetallics. The former is useful in improving the mechanical properties by Mg_2Si dispersion strengthening effect, and the latter is for the improvement of corrosion and wear resistance. The characteristics of MgSiX are described in details by the combination of grains-refining process and solid-state synthesis of Mg_2Si.

Grains Refining and Dispersion Strengthening Effects

Grains Refining Effect by Repeated Plastic Working

It is well known that the grains refinement improves the strength of materials. In this study, the repeated plastic working (RPW) process illustrated in figure is employed to produce the magnesium green compact with refined grains.

After feeding the raw powder mixture into the die installed in the press machine, it is consolidated by upper punch I. Upper punch II is put in the compact, and the plastic deformation of the green part by backward extrusion takes place in the die. After that, the compaction by punch I continues again.

One cycle in RPW process, which consists of compaction and backward extrusion by inserting two upper punches alternatively, requires about 6~10 s. The punch speed is about 0.5~1 m/s, and the energy consumed to mix and refine the raw material is controlled by the speed. The UTS remarkably increases according to the reduction of the grain size. It corresponds to Hall-Petch equation well. For example, when the number of cycles in RPW (N) is 200, the hot extruded AZ31 alloy shows the mean grain size of 4.3 μm and UTS of about 350 MPa, which is superior to the conventional one with 265 MPa UTS. Accordingly, RPW process is effective to refine magnesium grains for the improvement of the tensile properties.

Magnesium Composites with Mg_2Si Dispersoids

Mg_2Si intermetallic can be formed by the reaction of magnesium and silicon with an exothermic heat when pre-heating the green compact before hot extrusion. Silica (SiO_2) is also supplied as another raw material to synthesize Mg_2Si and MgO via the deoxidization by magnesium. These occur based on the following equations.

$$2Mg + Si \rightarrow Mg_2Si$$

$$4Mg + SiO_2 \rightarrow Mg_2Si + 2MgO$$

For example, when using AZ31 alloy powder with a mean particle size of 145 μm and SiO_2 powder with that of 22 μm, UTS of the composites with 8 mass% SiO_2 increases in proportion to the SiO_2 content, and is 362 MPa. The mean particle size of synthesized Mg_2Si /MgO is about 20 μm. It is the same as that of SiO_2 raw materials because the coarsening of the formed particles did not occur so much during solid-state reaction. Accordingly, from a viewpoint of the dispersion strengthening effect, the particle size control of Si and SiO_2 powder in the green compact is very important to produce the magnesium composites with fine Mg_2Si or Mg_2Si/MgO dispersoids.

Materials Design of MgSiX

The uniform distribution of in-situ formed Mg_2Si contributes to not only the improvement of mechanical properties but also the corrosion and wear resistance of the magnesium composite alloys.

Dependence of UTS on SiO_2 content via RPW process (N=200) compared to the conventional cold compaction process (N=1), when employing the elemental powder mixture of AZ31-5 mass%Si. In both composites, UTS proportionally increases in increasing the SiO_2 content due to the dispersion strengthening effect of fine Mg_2Si/MgO particles. The AZ31 composite via RPW, however, indicates higher tensile strength because of the matrix grains refinement by RPW.

Concerning corrosion resistance of the composite, when employing the elemental powder mixture of AZ31-2 mass%SiO_2 as raw materials. The current density is obviously reduced to 1/2~1/3 by including Mg_2Si, compared to the conventional AZ31 alloy. That is, the corrosion resistance of magnesium alloys can be improved by dispersing In-situ formed Mg_2Si particles in the matrix. Furthermore, the tribological properties are also remarkably improved. In particular, the friction coefficient under oil lubricant conditions is reduced by including MgO particles because of their "mild-offensive" properties to obstruct the abrasive phenomena on the counter materials. The addition of graphite particles is also effective in reducing stabilizing the friction coefficient due to their self-lubrication.

Recycling of Wasted SiO_2 Glass to Produce MgSiX

Considering that the main ingredient of the glass products is SiO_2, the glass scraps, in particular with a high purity, have a large possibility to be employed as input raw materials instead of silicon to form Mg_2Si via the deoxidization and reaction process. They are ground into fine powder, having a mean particle size of 16.8 μm. After mixing the SiO_2 glass particles with AZ31 powder, the elemental powder mixture is consolidated by cold pressing under the applied pressure of 600 MPa.

SiO_2 glass particle content of the powder mixture is 0, 2, 4, 6, and 8 mass%. After heating each green compact at 723 K for 300 s in nitrogen gas atmosphere to synthesize Mg_2Si/MgO, it is immediately consolidated into a full density by hot extrusion with an extruding ratio of 37. Furthermore, the Differential Scanning Calorimeter (DSC) curves of the green compacts heated at 723 K show no exothermic peak, which corresponds to the reaction between magnesium and SiO_2, when heating up to 910 K. It means that no glass particle exists in the green compact after heating at 723 K, and completely reacts with magnesium alloy powder during pre-heating.

The dependence of UTS, Yield Stress and elongation of hot extruded AZ31 composite alloys with Mg_2Si/MgO dispersoids on the glass content. It shows the gradual increase of tensile strength with increase in the SiO_2 content. This is because of the dispersion strengthening effect, when Mg_2Si/MgO fine particles are uniformly distributed in the matrix. The yield stress also increases according to the increase of SiO_2 content because of the extremely high Young's modulus of the dispersoids compared to the magnesium matrix. Elongation decreases with increase in the SiO_2 content, because in-situ formed Mg_2Si and MgO dispersoids are brittle compared to the AZ31 matrix. Accordingly, wasted SiO_2 glasses with a high purity are able to employ as raw materials to form Mg_2Si via the reaction and synthesis process suggested in this study.

Fabrication Methods for Magnesium Alloy

Shape Casting

Magnesium alloys can be cast into products by a variety of methods including high pressure die casting, low pressure permanent mould and sand casting, gravity permanent mould and sand casting, plaster/investrnent casting, as well as squeeze- and thixocasting. These methods cover a very wide range of processing parameters, emphasizing the need to tailor alloys and processes to obtain the required product performance. By far the most dominating casting method is high pressure die casting. During high pressure die casting, the mould, which is clamped together by hydraulic forces, is filled in a short time (10-100 ms) by moving the piston in the shot sleeve. After injection, a high pressure -150 bars) is exerted on the liquid metal to secure proper mould fig. The metal solidifies at high cooling rates (100-1000 K/s) leaving a fine Gained material with secondary dendrite arn spacing in the range 5-10 pm. For high performance casting of Mg-parts special consideration of the machine performance is needed. of particular importance is injection speed, and proper handling of the liquid alloy to secure satisfactory metal quality.

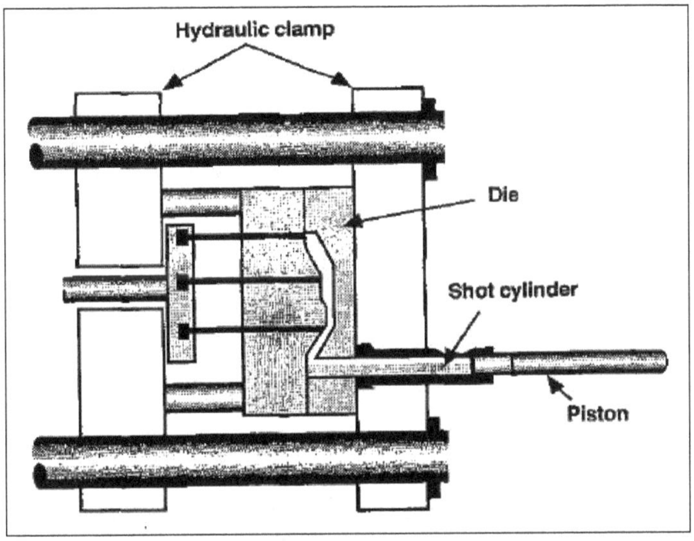

Schematic diagram of cold chamber high pressure die casting.

Wrought Applications

Due to the hexagonal structure, magnesium alloys must be formed plastically at temperatures exceeding 200 OC to allow the necessary number of slip systems to operate. Extrusion is usually carried out in two steps, involving pre-extrusion of large format billets into smaller diameter billets, which are subsequently extruded into the final shape. The specific pressure is higher than for the standard Al-alloys, and the extrusion speed is limited by hot shortness caused by local melting, a phenomena which increases m severity as the alloying content increases: Despite these disadvantages, Mg-extrusions are finding important uses for applications where the unique properties of Mg-alloys can be utilized. This includes aerospace applications, lightweight frames and racks, anodes for cathodic protection etc. Computer parts is a growing market due to light weight, high damping capacity for vibrations and excellent machinability. There is also a strong interest in Mg-extrusions for automotive applications as members of space frame constructions or light weight seats etc. Extrusions are frequently used as feedstock for forged applications.

An important development, which is expected to have a major impact on the competitiveness of Mg-extrusions, is the introduction of multistrand air-slip hot top casting of billets, figure. By this method, the two step extrusion process, will be replaced by a single step process, as commonly used for Al-extrusion.

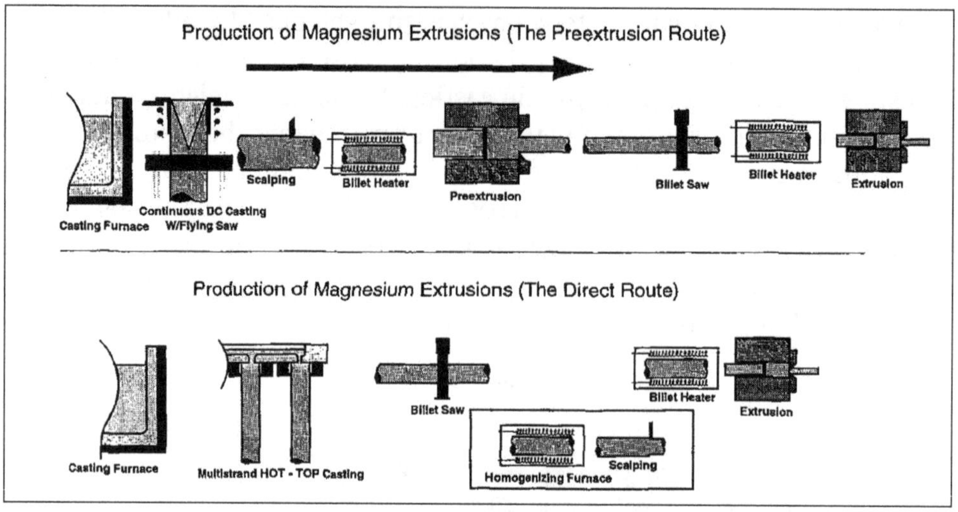

Alternative production routes for Mgextrusions.

Microstructure and Properties of Casting Magnesium Alloys

Due to chemical composition, casting magnesium alloys are divided into two groups. The first group includes alloys containing from 3 to 10% Al with the addition of Zn and Mn. They are characterized by a low cost of manufacture, good tensile strength, elongation, and resistance to atmospheric corrosion. The most popular representative of this group of alloys is AM50 alloy mainly for die casting and AZ91 that can be cast by sand and die casting method. The main advantage of these alloys is their relatively low price, while their disadvantage is low operating temperature—below 120°C. To increase the operating temperature of Mg-Al alloys, alloying elements are introduced, such as rare earth metals, strontium, calcium, and other. Application of rare earth elements to Mg-Al alloys causes a formation of stable thermodynamic phases at the grain boundaries, which provides motivation for improving creep resistance. In particular, the microstructure is connected by the $Al_{11}RE_3$ phase, which is characterized by a high thermodynamic stability to the temperature of ~180 °C. The high thermodynamic stability of this phase uses all the aluminum atoms to form this phase during spheroidization, which prevents the $Mg_{17}Al_{12}$ phase from forming. Consequently, these alloys may be applied in the automotive industry for engine elements, gearing, oil pans, and other structural materials working at temperatures of ~180 °C. High prices of rare earth elements and their small availability oblige us to search for alternative solutions. Some of those solutions are Mg-Al alloys to which strontium is added. The α-Mg solid solution and eutectic, which consists of one or more intermetallic phases, form the microstructure of Mg-Al-Sr alloys. The Al_4Sr phase is mainly observed in alloys with no more than 5% Al; however, the undesirable $Mg_{17}Al_{12}$ phase can be found in alloys with no more than 6% concentration of Al.

The second group includes alloys containing mainly Zn, RE, and Y, without the addition of Al, but always with the addition of Zr. These alloys can be used at a temperature higher than 120 °C (up to 250 °C), but the price of alloy additions increases the cost of their production. They are mainly used as sand castings. In this group, the most widely used alloys are WE54 and the latest EV31A

(Elektron 21). WE54 magnesium alloy reaches high specific strength, creep resistance, and corrosion resistance up to a temperature of 250 °C. The strength of this alloy is achieved essentially via precipitation strengthening. Depending on the aging temperature and time, the precipitating sequence in WE alloys has been reported to involve the formation of phases designated β″, β′, and β. EV31A is magnesium-based sand casting alloy containing neodymium, gadolinium, and zinc for used to approximately 200 °C. This alloy has high strength, good corrosion resistance, and excellent castability. In as-cast condition, EV31A alloy is characterized by a solid solution structure α-Mg with eutectic α-Mg + Mg_3(Nd, Gd) intermetallic phase on grain boundaries. Depending on the aging temperature and time, the decomposition of α-Mg supersaturated solid solution is as follows: α-Mg → β″ → β′ → β(Mg_3RE) → $Mg_{41}Nd_5$. EV31A is being used in both civil and military aircraft and also in the automobile (motorsport) industry.

Microstructure and Properties of Mg-Al Alloys

AM50 and AZ91 Alloys

Sand casting AM50 alloy is characterized by the structure of α-Mg solid solution with precipitates of two types of $Mg_{17}Al_{12}$ phases. First one of massive morphology, together with the solid solution, forms partially divorced eutectic (continuous $Mg_{17}Al_{12}$ + α-Mg) at the grain boundaries. Divorced eutectic $Mg_{17}Al_{12}$ + α-Mg is characterized by the presence of "islands" of α-Mg solid solution, solidifying due to eutectic reaction, which are surrounded by $Mg_{17}Al_{12}$ phase precipitates. The second one of plate morphology is created as a result of discontinuous diffusional transformation. Discontinuous precipitation occurs mostly in the α-Mg regions near the massive $Mg_{17}Al_{12}$ phase on the solid solution α-Mg grain boundaries in regions with higher aluminum. The volume fraction of α-Mg solid solution within $Mg_{17}Al_{12}$ phase precipitates is much smaller than it appears from the balance system. Moreover, globular precipitates of Al_8Mn_5 phase occur in the AM50 alloy.

The microstructure of Mg-Al alloys after sand casting, AM50 (a), AZ91 (b), SEM.

Casting magnesium alloy AZ91 is the most popular and relatively cheap in comparison with other magnesium alloys available on the market. Aluminum causes an increase in tensile strength and hardness of the alloy but only to a temperature of 120°C. The AZ91 magnesium alloy, like AM50 alloy in as-cast condition, was characterized by a solid solution α-Mg with discontinuous and continuous precipitates of the $Mg_{17}Al_{12}$ phase, with the difference that the volume fraction of the $Mg_{17}Al_{12}$ phase is higher in the AZ91 alloy. Moreover, the occurrence of Mg_2Si and Al_8Mn_5 phases has been provided.

After die casting, the structure of AM50 alloy is characterized by significant grain refining of the α-Mg solid solution; however, $Mg_{17}Al_{12}$ phase, together with the α-Mg solid solution, forms fully divorced eutectic on the grain boundaries of α-Mg solid solution.

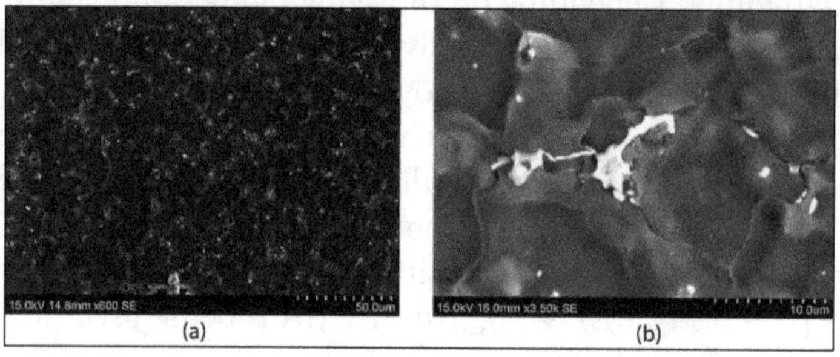

The microstructure of AM50 alloy after die-casting, SEM.

During annealing, precipitation of $Mg_{17}Al_{12}$ phase proceeds continuously and discontinuously. In sand casting, firstly, as a result of discontinuous precipitation, precipitates of plate $Mg_{17}Al_{12}$ phase are formed. The process is started on the grain boundaries of the α-Mg solid solution and consists of cellular growth of plate precipitation of $Mg_{17}Al_{12}$ phase in the direction of the central part of the solid solution grain. The growth of lamellar precipitates runs continuously until the alloy matrix reaches the equilibrium composition. The volume fraction of lamellar areas and the distance between the lamellas increases with the rise of temperature, aging time, and an aluminum content in the supersaturated areas. The second step is the coagulation of plate precipitates and the beginnings of continuous precipitation of $Mg_{17}Al_{12}$ phase in zones of increasing content of aluminum. Further annealing causes growth and coagulation of both types of precipitates. The extension of annealing time to 5000 h or increasing temperature of annealing to the 250 °C causes continuous growing and coagulation of precipitation of $Mg_{17}Al_{12}$ phase.

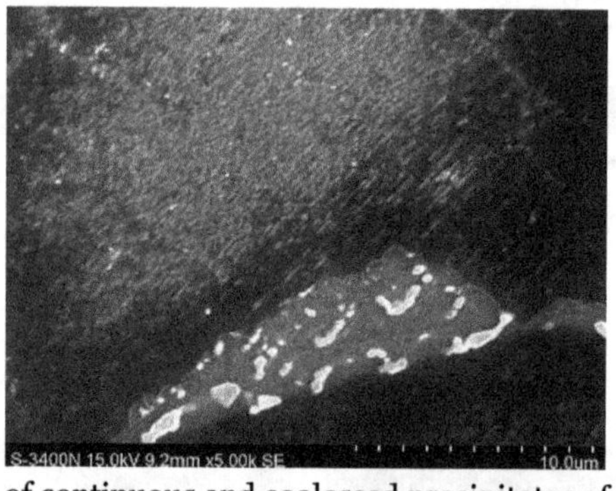

The zones of continuous and coalesced precipitates of Mg17Al12 phase in AM50 alloy, after annealing at 180 °C/500 h/air.

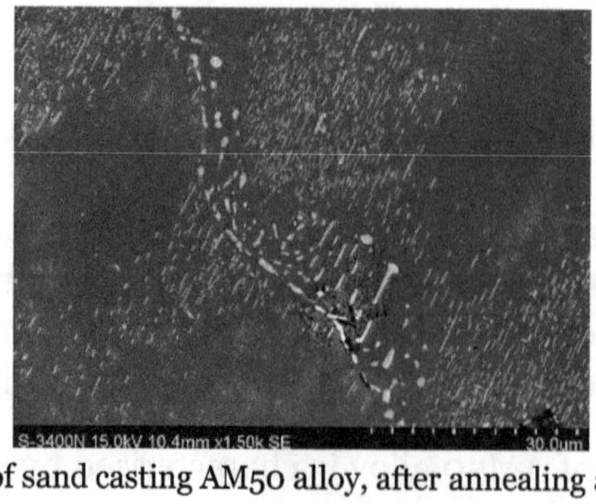

The microstructure of sand casting AM50 alloy, after annealing at 250 °C/5000 h/air.

The precipitation processes during long-term annealing in die casting AM50 magnesium alloy proceed similarly as in sand casting, but the difference is that discontinuous precipitation of $Mg_{17}Al_{12}$ phase is not observed in the first step of the process. After 4000 h of annealing, a big, coagulated precipitation of $Mg_{17}Al_{12}$ phase makes the structure of the alloy on the grain boundaries of α-Mg solid solution. Long-term annealing at 250°C temperature causes growth and coagulation of continuous precipitates of $Mg_{17}Al_{12}$ phase.

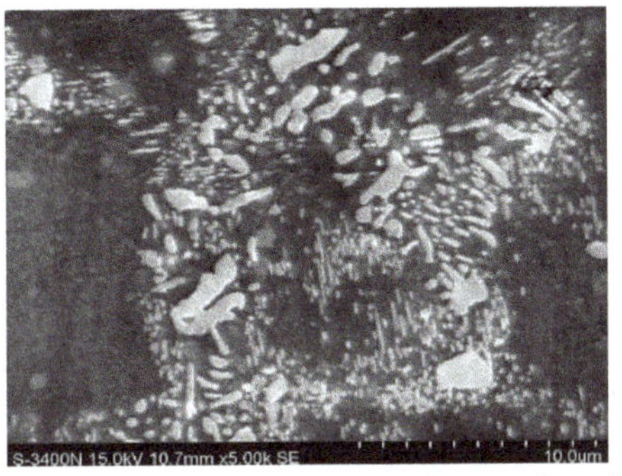

Coalesced continuous precipitates of Mg17Al12 phase in die-casting AM50 alloy after annealing at 180 °C/4000 h/air.

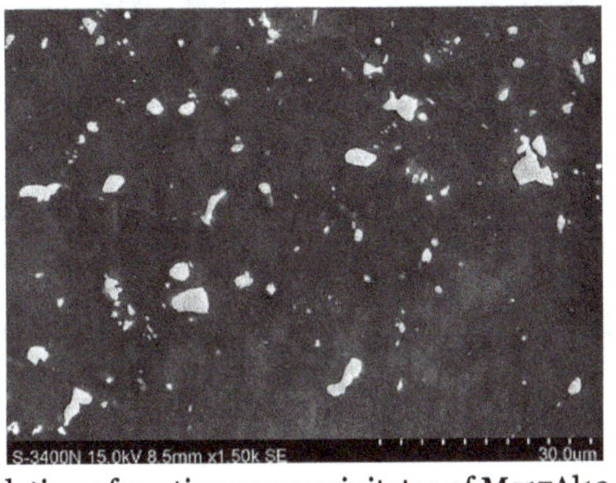

Coagulation of continuous precipitates of Mg17Al12 phase in die-casting AM50 alloy after annealing at 250 °C/4000 h/air.

The precipitation processes of the $Mg_{17}Al_{12}$ phase during long-term annealing of sand casts resulted in the growth of the hardness of the alloy alongside with lengthening the annealing time. However, in die casts, the structure of the alloy undergoes degradation in the result of precipitation and the coagulation of $Mg_{17}Al_{12}$ phase, what reduces hardness significantly.

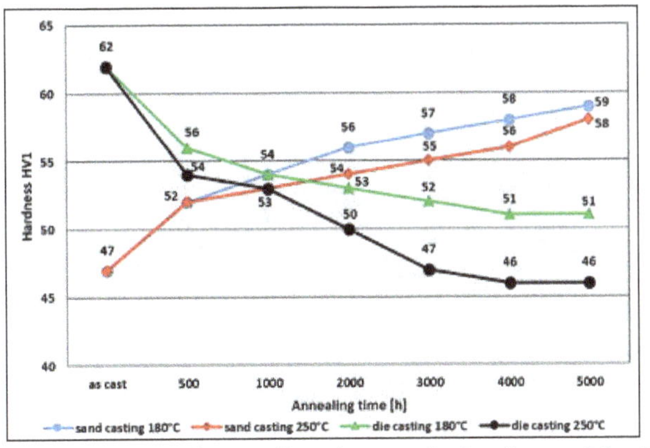

Influence of temperature and annealing time on the hardness of the AM50 alloy.

AM50 and AZ91 alloys due to very good mechanical properties at an ambient temperature and low at elevated temperature can only be used at up to 120 °C. This is connected with the presence of $Mg_{17}Al_{12}$ phase, which helps to increase the tensile strength at an ambient temperature but decreases the mechanical properties at an elevated temperature.

Microstructure and Properties of Mg-Al-X (RE, Sr) Alloys

AE44 Alloy

Addition of rare earth elements (mainly cerium, lanthanum, and neodymium) to an alloy with 4% aluminum increases durability and operating temperature of the alloy. The microstructure of the AE44 alloy after sand casting is dominated by the $Al_{11}RE_3$ phase. Moreover, the Al_2RE and $Al_{10}RE_2Mn_7$ phases occur. Die-casting causes grain refining of the solid solution. Diverse morphology of the interdendritic phases occurs at the α-Mg grain boundaries. Needle precipitates of the $Al_{11}RE_3$ phase together with the α-Mg solid solution form a eutectic like–fiber morphology. In contrast to the situation for sand casting, the metastable $Al_{2.12}RE_{0.88}$ phase forms next to eutectic areas.

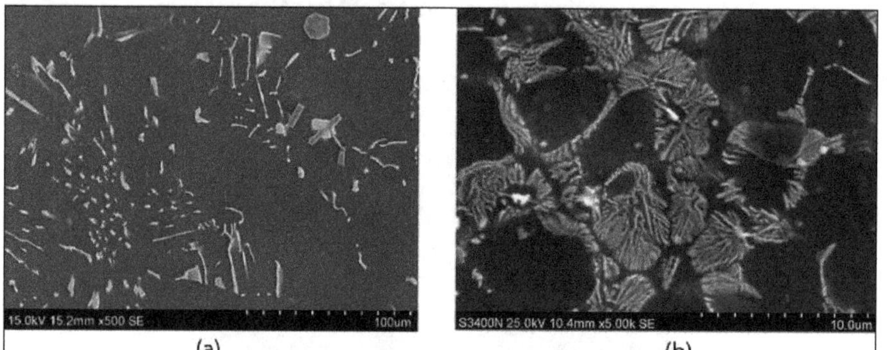

(a) (b)

The microstructure of the AE44 alloy after sand casting (a), die casting (b), SEM.

The $Al_{11}RE_3$ and Al_2RE compounds are typical in Mg-Al-RE alloys. The $Al_{2.12}RE_{0.88}$ is not an equilibrium phase in the binary Al-RE (Al-La) system. The presence of this compound is a result of rapid crystallization during die casting and that it is a metastable phase at room temperature. The $Al_{2.12}RE_{0.88}$ compound was not found in slowly cooled sand cast AE44 alloy.

Precipitates of Mg17Al12 phase in the sand cast AE44 alloy after annealing at 180 °C/3000 h.

After long-term annealing at 180 °C for 3000 h, the structure of AE44 alloy reveals no significant changes. Continuous precipitation of the $Mg_{17}Al_{12}$ phase is observed only in aluminum supersaturated areas of the α-Mg solid solution in the sand cast alloy. The morphology of $Mg_{17}Al_{12}$ phase suggests continuous precipitation from a supersaturated solid solution. These precipitates were occasionally observed in the alloy. Increasing the annealing temperature to 250 °C accelerates the process of fragmentation and spheroidization of $Al_{11}RE_3$ phase precipitates.

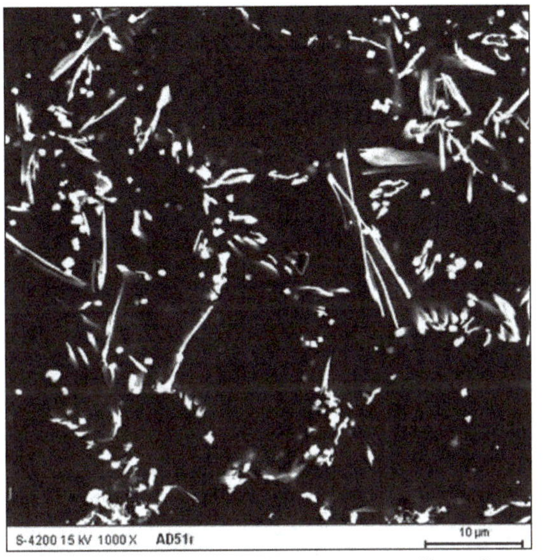

Fragmentation of Al11RE3 phase precipitates in the die-cast AE44 alloy after annealing at 250 °C/3000 h.

During long-term annealing of die-cast AE44 alloy, the metastable $Al_{2.12}RE_{0.88}$ phase undergoes a transition into the equilibrium Al_2RE phase.

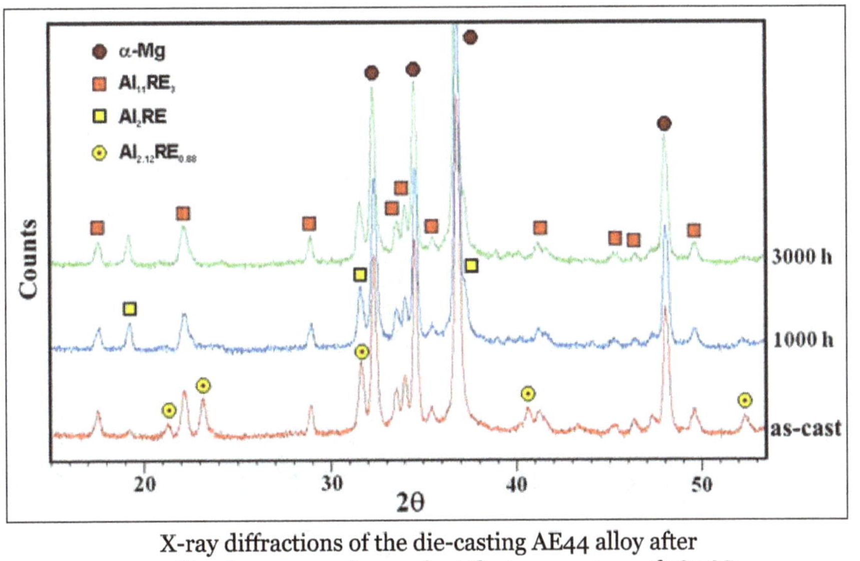

X-ray diffractions of the die-casting AE44 alloy after annealing for 1000 and 3000 h at the temperature of 180 °C.

An increase in the hardness (only sand cast) after annealing at the temperature of 180 °C is caused by the separation of $Mg_{17}Al_{12}$ phase in the areas of increased aluminum content. The fragmentation and spheroidization of $Al_{11}RE_3$ phase after annealing at temperatures of 180 and 250 °C caused a significant decrease in hardness and tensile strength of die casting AE44.

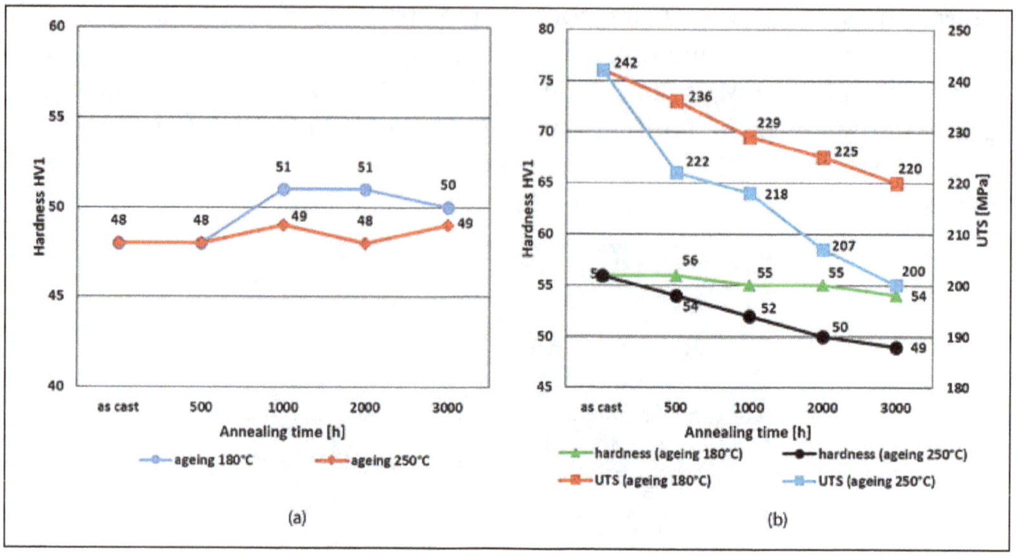

The influence of temperature and annealing time on the hardness and tensile strength of the AE44 alloy after sand casting (a) and die casting (b).

AJ62 Alloy

The AJ62 alloy is characterized by the structure of the α-Mg solid solution with precipitates of intermetallic phases of type: (Al, Mg)$_4$Sr, Al$_3$Mg$_{13}$Sr, and Mn$_5$Al$_8$. The (Al, Mg)$_4$Sr phase and the solid solution form eutectic of α-Mg + (Al, Mg)$_4$Sr. However, the Al$_3$Mg$_{13}$Sr phase occurs at the grain boundaries of the α-Mg solid solution in the direct surroundings of eutectic areas. Globular precipitates of the Mn$_5$Al$_8$ phase occur at grain boundaries and inside the grains of the α-Mg solid solution. In the thin-wall die casts, eutectic α-Mg + (Al, Mg)$_4$Sr occurs at the grain boundaries. Moreover, for thick-wall casting, which is characterized by a lower cooling rate, the Al$_3$Mg$_{13}$Sr phase occurs.

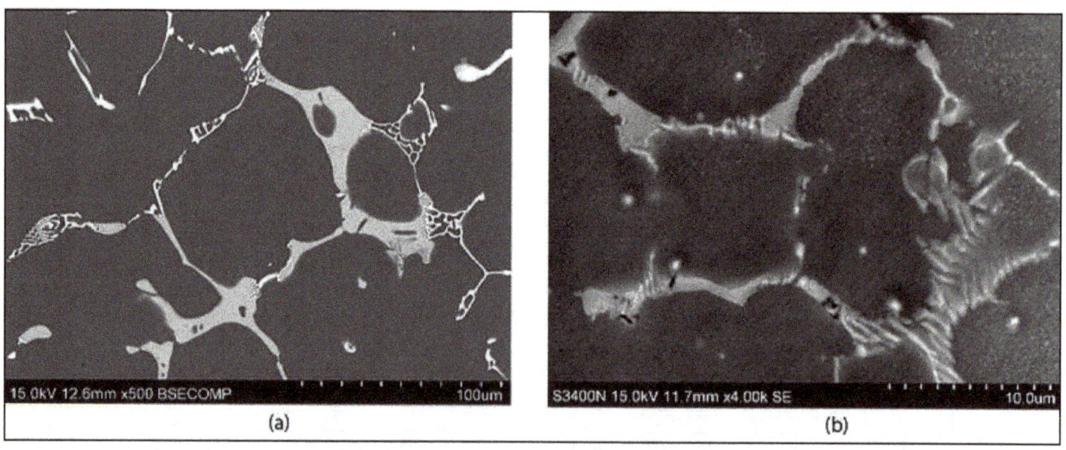

The microstructure of AJ62 alloy after sand casting (a), die casting (b), SEM.

The applied casting technology has no influence on the type of changes in the AJ62 alloy microstructure during long-term annealing. The first step occurs at ~180 °C (250 °C/500 h) and is characterized by precipitation of the Mg$_{17}$Al$_{12}$ phase in areas of higher aluminum volume (as in AE44 alloy) and the beginning of Al$_3$Mg$_{13}$Sr phase decomposition.

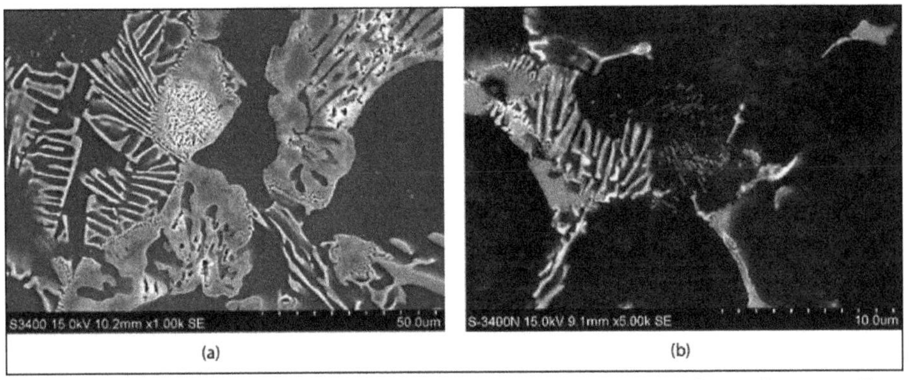

Precipitates of Mg17Al12 and initial decomposition of Al3Mg13Sr phase after annealing of sand casting for 500 h at 250 °C (a) and die casting for 4000 h at 180 °C (b).

Decomposition of this phase results in the formation of plate precipitates of the $(Al, Mg)_4Sr$ phase separated by areas of α-Mg solid solution. The second step occurs at ~250 °C and consists of complete degradation of the $Al_3Mg_{13}Sr$ phase by the reaction $Al_3Mg_{13}Sr \rightarrow (Al, Mg)_4Sr + \alpha\text{-}Mg$. The third step occurs above 300 °C. In the α-Mg solid solution, there are only precipitates of the primary and secondary (generated from $Al_3Mg_{13}Sr$ phase decomposition) $(Al, Mg)_4Sr$ phase. Long-term annealing at this temperature leads to fragmentation and coagulation of $(Al, Mg)_4Sr$ phase precipitates.

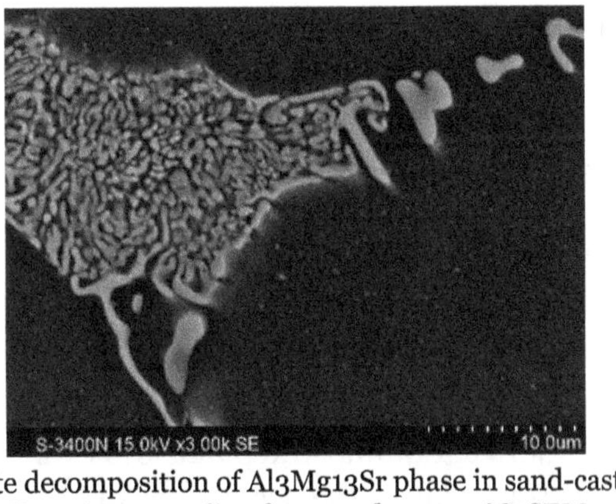

Complete decomposition of Al3Mg13Sr phase in sand-cast AJ62 alloy after annealing for 1000 h at 350 °C, SEM.

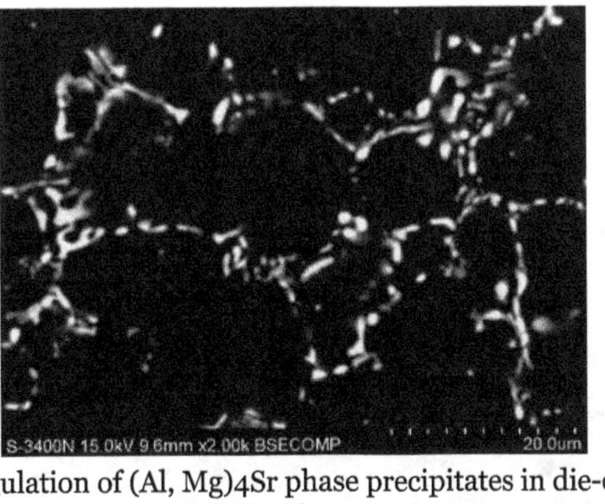

Coagulation of (Al, Mg)4Sr phase precipitates in die-cast AJ62 alloy after annealing for 4000 h at 350 °C, SEM.

The hardness and tensile strength evolution of AJ62 alloy as a function of aging time for isothermal aging at 200, 250, and 350 °C is shown in figure.

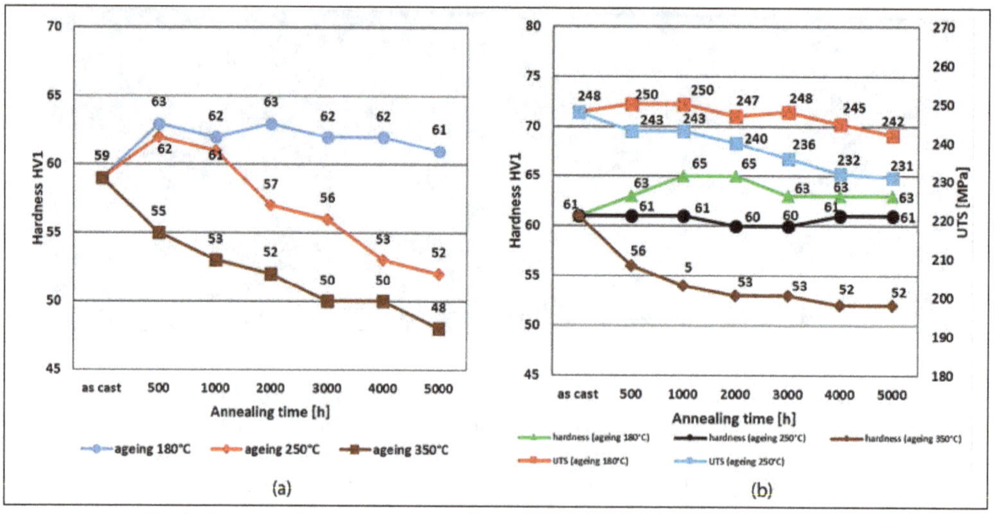

The influence of temperature and annealing time on the hardness and tensile strength of the AJ62 alloy after sand casting (a) and die casting (b).

Increase in hardness after annealing at the temperature of 180 °C as well as in the first stage of annealing at the temperature of 250 °C (only sand cast) is caused by the separation of $Mg_{17}Al_{12}$ phase in the areas of increased aluminum content. The hardness of the die cast, unlike of sand cast, does not decrease after annealing at the temperature of 250 °C. It is related to the lower number of $Al_3Mg_{13}Sr$ phase precipitates. The decomposition of $Al_3Mg_{13}Sr$ phase and coagulation of $(Al, Mg)_4Sr$ phase precipitates after annealing at the temperature of 350 °C cause a significant decrease in hardness and tensile strength of AJ62 alloy, regardless of applied casting technology.

Microstructure and Properties of Mg-Zr-X Alloys

WE54 Alloy

The WE54 alloy in as-cast condition was characterized by a solid solution structure α-Mg with eutectic α-Mg + β on grain boundaries. Equilibrium β phase is isomorphic to the Mg_5Gd phase and is identified as a $Mg_{14}Nd_2Y$ phase. Moreover, the occurrence of MgY, Mg_2Y, and $Mg_{24}Y_5$ phases has been provided. Also, the zirconium-rich core areas have been observed. Zirconium-rich core areas are ellipsoidal or nearly circular.

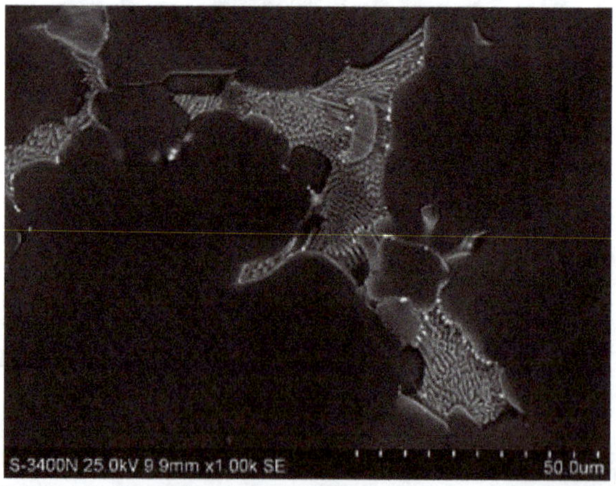

The WE54 alloy microstructure in as-cast condition, SEM.

Zirconium-rich core areas in the WE54 alloy.

The high tensile strength of WE54 alloy is related to dispersive precipitations in the microstructure. After 4 hours of aging at 250 °C two types of metastable phases, coherent or semi-coherent with the matrix were observed. The first type of precipitates is, coherent with the matrix, β″ phase, characterized by type DO_{19} structure. The second phase is, semi-coherent with the matrix β′ phase, characterized by an orthorhombic, space-centered structure. Extension of the aging time up to 16 hours (T6 treatment) causes the disappearance of β″ phase. The β′ phase grows and changes its shape from spherical to lamellar one. Extension of the aging time up to 48 h causes the growth and the change of shape of the β′ phase precipitates from spheroidal into lamellar and the beginnings of the phase transformation $\beta' \rightarrow \beta_1$. The resulting new phase, identified as β_1 phase, nucleates heterogeneously on the β′ phase precipitates. The β_1 phase is characterized by a face-centered cubic structure. Further aging (96 h) increases the amount of β_1 phase and causes the emergence of equilibrium β phase. Precipitation processes end with the emergence of the equilibrium β phase. It is formed during aging for a long time at 250 °C and above.

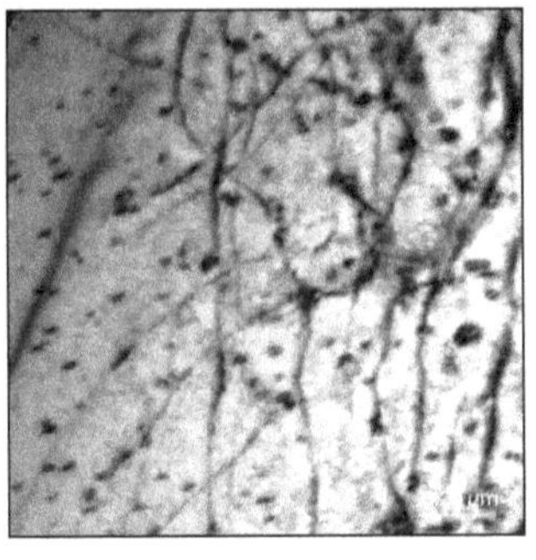

Fine-dispersed precipitates of β″ and β′ phases in the WE54 alloy after aging at 250 °C/4 h, TEM.

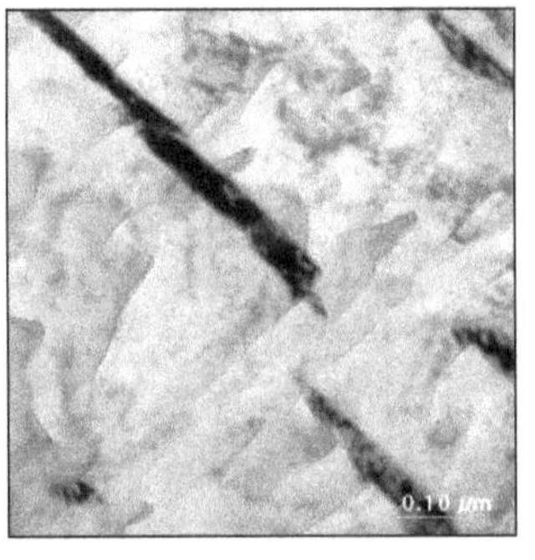

β′ phase precipitates in the WE54 alloy after aging at 250 °C/16 h, TEM.

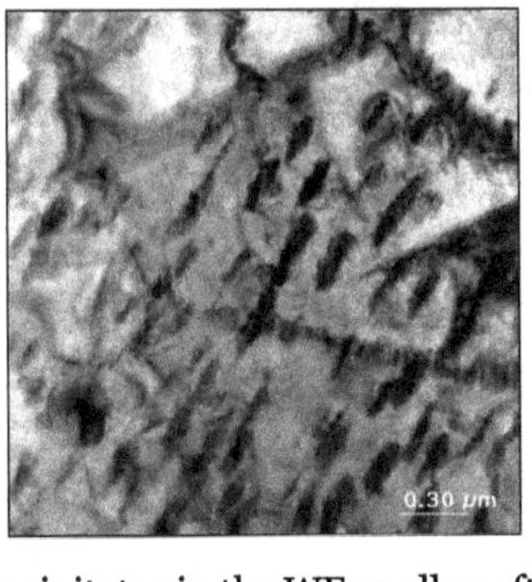

β′ phase precipitates in the WE54 alloy after aging at 250 °C/48 h, TEM.

β′ and β1 phases in the WE54 alloy after aging at 250 °C/96 h, TEM.

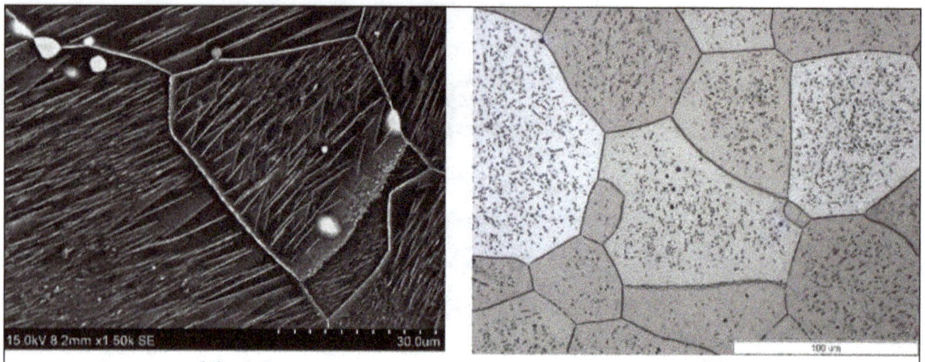

(a) (b)

Equilibrium β phase in (a) the WE54 alloy after aging at 300°C/96 h—SEM and (b) 420 °C/22 h—LM.

The evolution of hardness and tensile strength as a function of aging time for isothermal aging at 200 and 250 °C is shown in figure. In practice, for this alloy, aging at 250 °C for 16 h is applied (T6 treatment), allowing to obtain maximum tensile strength, which is related to the presence of β″ and β′ phases precipitates in the structure.

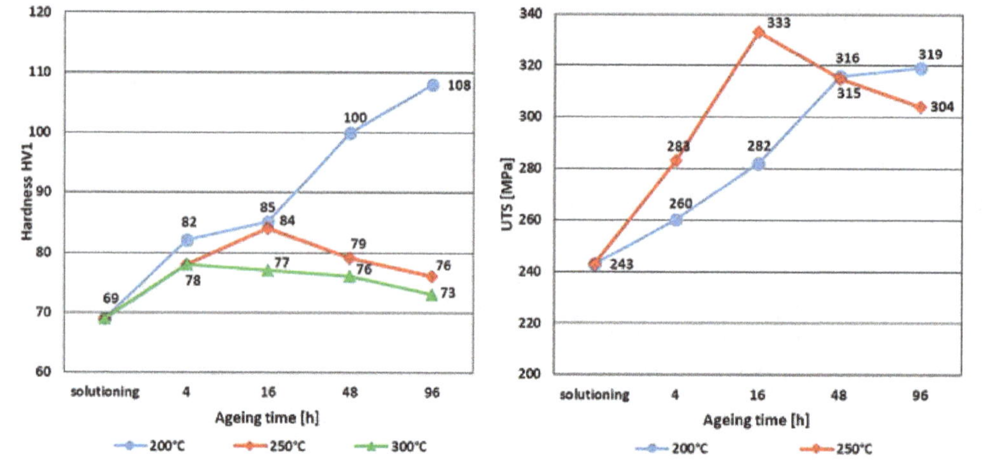

(a) (b)

The influence of temperature and annealing time on the hardness (a) and tensile strength (b) of the WE54 alloy.

The Mg-Y-Zr (WE43 and WE54) alloys are widely used in aircraft and automotive industries. Due to the presence of yttrium, the Mg-Y-Zr alloy belongs to the most expensive of magnesium alloys. This is the main reason for searching other alloys that will fulfill the performance requirements at lower manufacturing costs. One of them is EV31A (Elektron 21) alloy.

EV31A (Elektron 21) Alloy

The EV31A alloy is characterized by the α-Mg solid solution structure with eutectic α-Mg + Mg_3(Nd, Gd) on the grain boundaries and regularly shaped precipitates of a Mg_3Gd phase. The Mg_3(Nd, Gd) phase is a modification of Mg_3Nd phase with neodymium substituted by gadolinium without destroying the crystal structure, due to the reasonably small difference in the atomic radii between gadolinium r_{Gd} = 0.1802 nm and neodymium r_{Nd} = 0.1821 nm. Phases occurring in the EV31A alloy take the forms $(Mg, Zn)_3(Nd, Gd)$ and $Mg(Gd, Nd)_3$.

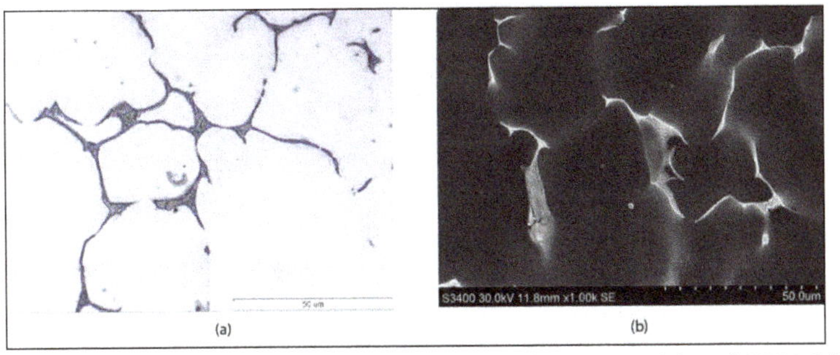

The EV31A alloy microstructure in as-cast condition, LM (a) and SEM (b).

Similarly, as in the WE54 alloy, high tensile strength is connected with dispersive precipitates produced during aging. In the first stage of the aging process (200 °C/4 h and 16 h), the product of the α-Mg decomposition is fully coherent and semi-coherent with the matrix precipitates of β″ and β′ phases, respectively.

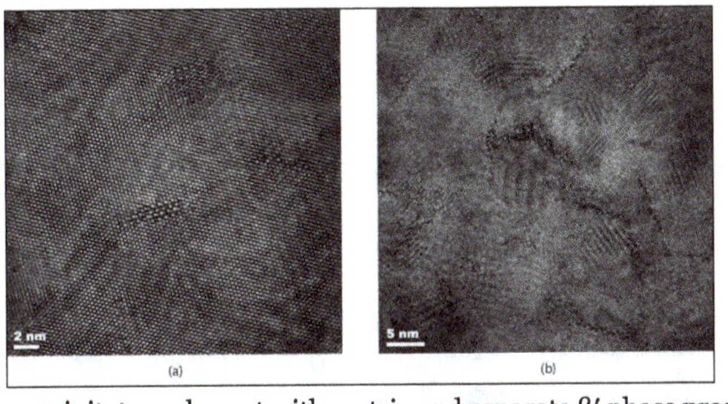

β″ phase precipitates coherent with matrix and separate β′ phase precipitates in EV31A alloy after aging at 200 °C/4 h (a) and 200 °C/14 h (b), HRTEM image.

The β″ phase is characterized by the lattice type DO19 (a = 0.64 nm, c = 0.52 nm), while β′ has a face-centered cubic structure (a = 0.72 nm). The second stage (200 °C/48H) is the formation of a stable equilibrium β(Mg_3Nd) phase. The β phase is non-coherent with the α-Mg matrix. The β phase precipitation process is similar to $β_1$ phase precipitation in the WE54 alloy. In the immediate vicinity of the β′ phase, lamellar precipitates of the equilibrium β phase arise. This indicates that the resulting new phase nucleates heterogeneously at the β′ phase precipitates. After 96 h of aging, the volume fraction of equilibrium β phase increases. The β phase is incoherent with the matrix and identified as an Mg_3Nd phase with face-centered cubic structure (a = 0.7410 nm). On the other hand, aging of this alloy at 300 °C leads to the appearance in the structure, next to the lamellar β phase precipitates, of $Mg_{41}Nd_5$ phase precipitates at the solid solution grain boundaries.

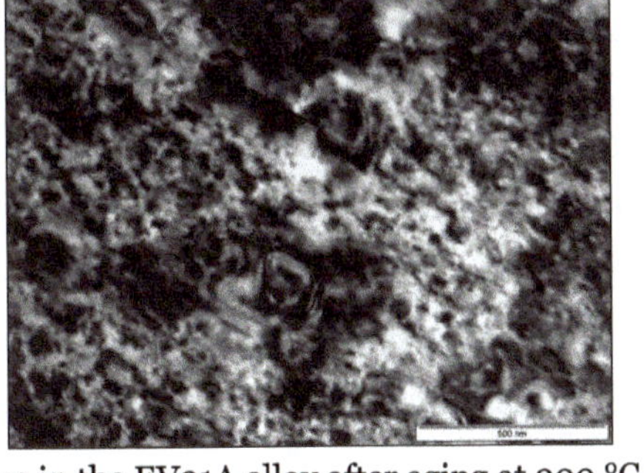

β′ and β phases in the EV31A alloy after aging at 200 °C/48 h, TEM.

β and Mg41Nd5 phases in the EV31A alloy after aging at 250 °C/48 h, TEM.

In the EV31A alloy, after long-term aging at a temperature 350 °C/500–5000 h/air, there are no precipitates of equilibrium β phase.

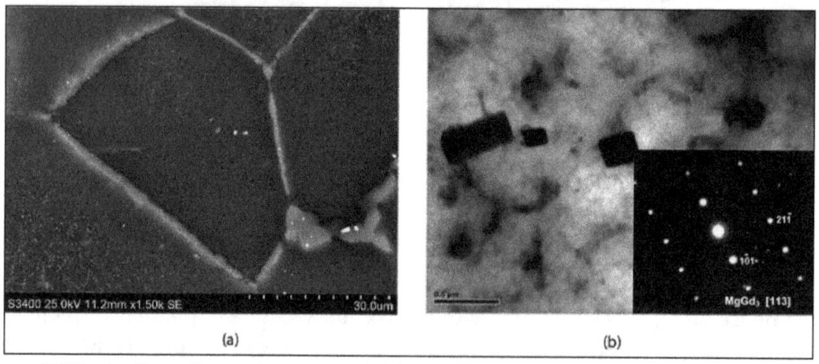

The microstructure of EV31A alloy after aging at 350 °C/1000 h: (a) Mg41Nd5 phase precipitates network at α-Mg solid solution grain boundaries, SEM; (b) TEM image and electron diffraction pattern of MgGd3 phase.

$Mg_{41}Nd_5$ phase creates a characteristic network at the grain boundaries of the α-Mg solid solution. In its vicinity, there are regular $Mg(Nd, Gd)_3$ phase precipitates.

Generally, the investigated alloy showed that the decomposition of α-Mg supersaturated solid solution with increasing aging time is as follows:

$$\alpha-\mathrm{Mg} \rightarrow \beta'' \rightarrow \beta' \rightarrow (\mathrm{Mg_3\,RE}) \rightarrow \mathrm{Mg_{41}\,Nd_5}.$$

The evolution of hardness and tensile strength as a function of aging time for isothermal aging at 200, 250, and 300 °C is shown in figure. Alloy showed a remarkable hardening at 200 °C temperature, and the peak hardness and good tensile strength can achieve after aging 200 °C/16 h due to the precipitation of β'' and β' phases. The peak hardness was shortened with an increase of the aging temperature. There was not any peak hardness in case of an alloy aged at 300 °C. The hardness of EV31A alloy decreases to 45 HV after 500 h of annealing due to precipitation of β and $Mg_{41}Nd_5$ phases.

In practice, for this alloy, aging at 200 °C for 16 h is applied (T6 treatment), allowing to obtain maximum tensile strength, which is related to the presence of β'' and β' phases precipitates in the structure.

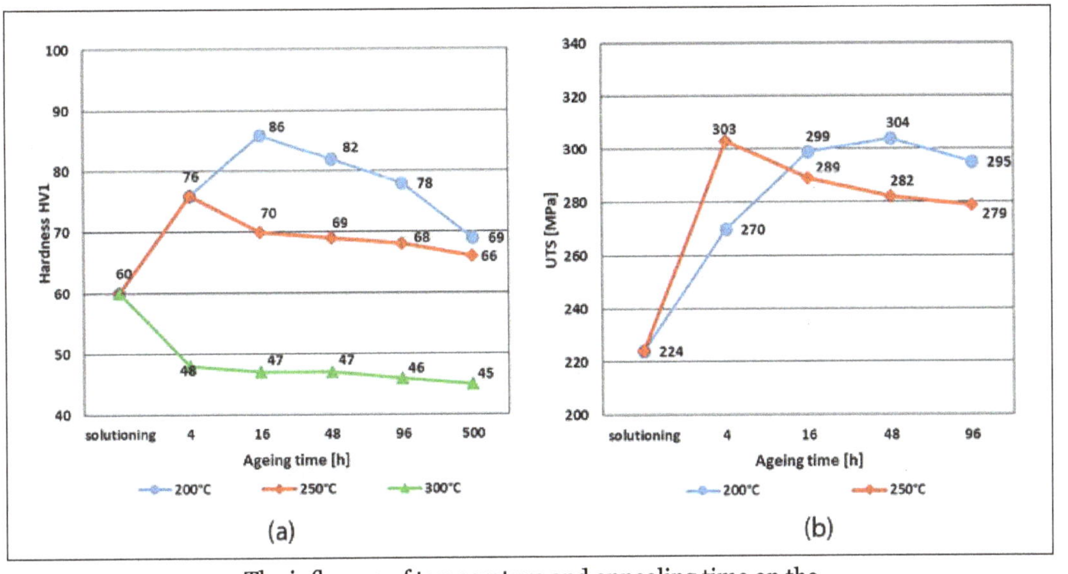

The influence of temperature and annealing time on the hardness (a) and tensile strength (b) of the EV31A alloy.

Corrosion Types of Magnesium Alloys

Magnesium (Mg) alloys are susceptible to corrosion in aggressive environments. Corrosion of Mg alloys depends greatly on their composition and microstructure (grain size, the size, shape and distribution of second phases), post-processing and media. In most cases, localized corrosion, such as pitting corrosion and filiform corrosion, generally occurs due to microgalvanic corrosion between the intermetallic compounds and their neighboring α-Mg matrix.

Magnesium (Mg) and its alloys can find extensive applications in transportation, 3C (computer, communication and consumer electronics) products and biomedical fields due to the lightweight, high specific strength and good biocompatibility. Mg alloys are susceptible to deterioration in an aggressive solution or harsh environment due to their low corrosion potential. Like other metals, corrosion modes of Mg alloys can be divided into uniform or general corrosion and localized corrosion based on the phenomenon from electrochemical, composition and microstructural perspectives on macroscale. The latter predominately includes galvanic corrosion, pitting corrosion, filiform corrosion, intergranular corrosion (IGC), exfoliation corrosion (EFC), crevice corrosion, stress corrosion cracking (SCC), corrosion fatigue (CF) and erosion corrosion. EFC and crevice corrosion, however, have been rarely reported.

Corrosion form is fundamentally influenced by metallurgical factors such as chemical composition, grain size and shapes, size, shape and distribution of secondary phases or intermetallic compound particles, inclusions, solute-aggregated grain boundaries (GBs), crystallographic orientations and dislocation density. Moreover, post-processing (i.e., extrusion and rolling) and post heat-treatment (i.e., T4, T5 and T6) exert a significant impact on corrosion mechanism. These treatments can result in evident changes in microstructure and stress. Furthermore, corrosion mode is also highly related to chlorine ion content and pH value of the corrosive solution. Localized corrosion is prone to occur under low pH value or acidic and neutral solution.

In addition, modern characterization techniques such as environmental scanning electron microscopy (ESEM), scanning electrochemical microscopy (SECM) and Kelvin probe force microscopy (KPFM) open a door for understanding corrosion mechanisms of Mg alloys at a microscale level.

Investigation with ESEM shows that pitting corrosion of Mg alloy AZ91 merely emerges at the sites (AlMn phases and scratches) with segregation of salt particles. Application of SECM reveals that pitting corrosion of Mg is initiated from film-free region. The concentration of Cl^- ions and composition of the surface film dominate the incubation and growth of active sites in film-free region. Higher concentrated chloride ions result in rapid initiation and propagation of active spots.

It is critical for design engineers to have an insight into the corrosion mechanisms and morphology of Mg alloys and to avoid sudden pre-failure of Mg parts.

Forms of Corrosion

Uniform Corrosion

Uniform corrosion can be observed on solution (T4)-treated and peak-aged WE43 alloy in 3.5 wt.% NaCl solution saturated with $Mg(OH)_2$. This scenario for the T4-treated Mg alloy relates to the formation of a protective corrosion product layer, which is composed of an inner MgO layer and an outer $Mg(OH)_2$ layer. The improvement in corrosion resistance for the aging treated alloy is ascribed to the interactions between the finely distributed precipitates and the growth of corrosion reactions.

Galvanic Corrosion

Galvanic corrosion is a common corrosion feature of Mg alloys. For example, galvanic corrosion may occur when Mg parts are connected or welded with the steels in lightweight powertrains and automotive bodies. In human body fluids, severe galvanic corrosion emerges when degradable Mg screws are connected with the bone plates made of titanium and stainless steel. This scenario is attributed to Mg that is located in the most active end in galvanic series of metals. Therefore, Mg and its alloys become the anode when connected to cathodic metals such as stainless steels and titanium alloys. This kind of galvanic corrosion is also termed bimetallic corrosion.

In some cases, galvanic corrosion occurs between two layers with different grain size in the same alloy. For instance, hot extruded Mg alloy ZK60 shows coarse grains in the outer layer and fine grains in the interior or middle layer. The outer layer with a higher open corrosion potential (OCP) is passivated, while the middle layer with a lower OCP is preferentially corroded. It is the initial OCP rather than grain size that decides the initiation of galvanic corrosion.

At a micro-scale level, micro-galvanic corrosion may occur between the α-Mg matrix with a lower OCP and its surrounding second phases or intermetallic compounds with a higher OCP in Mg alloys. The matrix is, thus, anodic and preferentially corroded. This theory is used to explain the mechanism of localized corrosion, e.g., pitting corrosion, IGC and EFC.

The nobler metals have a direct influence on the corrosion degree of the connected Mg alloys. The potential difference between Mg and other metals will be different, which can generate disparate galvanic couples and galvanic current. The galvanic current increases with the potential difference, that is, the bigger potential difference between Mg and dissimilar metal, the easier for Mg subject to corrode. The intensity of corrosion depends on the relative Volta potential differences of the micro-constituents, as well as their amounts, compositions and distributions within the alloys by KPFM.

The variations of pH may change the electrode reaction and polarity of galvanic metals. For instance, in an Al-Mg galvanic coupling in a neutral or weakly alkaline NaCl solution, Al acts as cathode. But with the dissolution of Mg, the solution changes into alkaline and aluminum turns into anode, which is the alkalinized effect of corrosion products of Mg alloys. Therefore, in an alkaline medium, that is, pH > 8.5, Mg alloys are basically non-corrosive. Mg alloys in acidic and neutral solutions, except for chromic acid, phosphoric acid and hydrofluoric acid that can form passive film, are susceptible to corrosion. When pH > 10.5, due to the formation of a protective $Mg(OH)_2$ film, Mg alloys have certain corrosion resistance. However, if a solution contains Cl^-, corrosion rate of Mg alloys will increase greatly.

Geometry factors such as cathode/anode area ratio, insulation distance between cathode and anode, depth of the dissolved film covering the galvanic couples and interactive mode of interaction caused by galvanic coupling can also affect the corrosion of Mg alloys.

The distance between galvanic couples is also important for corrosion behavior of Mg alloys. The distance of the galvanic current of the corrosion behavior of Mg alloys is limited. Corrosion generally occurs near the edge of such a specimen. The declining amplitude decreases with the increase in the distance similarly to an exponential decay.

Pitting Corrosion

Pitting corrosion is one of the most common corrosion types of Mg alloy. The generation of pits is related to chemical composition and specific microstructural features at the exposed Mg alloy surface. Chemical composition, grain size, heat treatment condition, morphology and distribution of the second phases all have an important influence on the corrosion of Mg alloys.

For sand-cast Mg alloy ZE41-T5 immersed in aggressive environment, pits initiate in the α-Mg grains neighboring T-phase (Mg_7Zn_3RE) at grain boundaries (GBs). This is attributed to galvanic corrosion, which is driven by the potential difference between the T-phase and α-Mg phase. In a subsequent immersion, the Zr-rich zones located in the center of grains become preferential corrosion sites. Small pits appear in the zones and severe attack develops.

In most cases, second phases that is, $Mg_{17}Al_{12}$ and Al_8Mn_5 are cathodic relative to α-Mg matrix, whereas in very limited cases, the second phases, containing rare earth (RE) and Mg in Mg alloy GW93 (Mg-Y-Gd), are more active than Mg matrix.

The microstructure constituents of Mg alloys are usually characterized by the α-Mg matrix and second phase or intermetallic phases, for instance, β ($Mg_{17}Al_{12}$), $MgSi_2$, AlMn(Fe), MgZn, Mg_2Zn, Mg_2Ca, Mg_2Sn, Mg_2Cu, Al_8Mn_5, I-phase (Mg_3Zn_6Y) and T-phase. Because in most cases, the electrode potential of these second phases is much higher than that of the α-Mg substrate, dissolution tends to emerge in the α-Mg matrix around second-phase particles. The presence of small Mg_2Ca particles coexisted with Fe and Si results in pitting corrosion of Mg-Ca alloys.

Occasionally, pitting corrosion is competitive when a number of second phases exist in Mg alloys.

In extruded Mg alloy AZ80, there may be three kinds of intermetallic compounds: $Mg_{17}Al_{12}$, Al_8Mn_5 and Mg_2Si. Voltaic potential differences between them and the α-Mg matrix phase or contact potential difference are not the same. And corrosion preferentially occurs around AlMnFe second phase due to the fact that the potential difference between the AlMnFe phase and matrix is the biggest.

For Mg-Al alloys, pitting corrosion occurs along the net structure of $Mg_{17}Al_{12}$ selectively. TGC will take place in die casting Mg alloys AM50 and AM20 caused by coring. At first, corrosion pits were formed in Mn-bearing particles, then developed into corrosion cracks and extended and terminated in the high Al region or GBs of the crystals.

Pitting corrosion can be affected by grain size. Refined grains make the corrosion morphology from pitting to homogeneous or uniform corrosion.

The microstructure of an extruded Mg alloy AM60 is composed of α-Mg matrix, β phase along GBs and granular AlMn phase in grains. The Al/Mn atomic ratio in AlMn phase is 1.12–1.15:1. In addition, there are a small number of Mg, Si elements and a small amount of deformation twins. The GB mainly consists of β phase. The AlMn phase is major in the grain that forms a cubic shape. There are both fine and coarse grains in the microstructure. Simultaneously, recrystallized grains form in the coarse grains. This result is due to the hot extrusion processing. The thick plate has a faster cooling rate for the outer layer and relatively slower cooling rate in the interiors, so localized recrystallization and grain abnormal growth phenomenon may occur.

In addition, pH value has a great influence on corrosion morphology of Mg alloys. In acidic and neutral NaCl solutions, the α-Mg matrix phase of Mg alloy AM60 around AlMn particles is preferential to corrode. Thus, pitting corrosion of Mg alloy AM60 occurs. This may be the result of local galvanic corrosion due to a higher voltage of the AlMn phase than α-Mg matrix and β phase. In alkaline solutions, uniform corrosion of high alumina regions is observed with cellular corrosion morphology. At pH 12, the sample surface has only a few shallow corrosion pits, which is well protected. From the morphology of the pits, the site of incipient pitting corrosion is related to the AlMn phase. The pits form on α-Mg in the area adjacent to the AlMn phase and there are many cracks in the matrix near the pits. The β phase does not play a role in the corrosion process of the extruded AM60 alloy.

Figure illustrates the pitting corrosion mechanism of extruded AM60 alloy. The oxide film formed on Mg alloy surface is discontinuous, which provides the adsorption sites for anions such as Cl^-. Here, the corrosion pits of Mg alloys are hemispherical. Cl^- ions are adsorbed in α phase adjacent to the AlMn particles. At the self-corrosion potential (−1.53 V/SCE), the passivation film is broken, the α phase begins to dissolve and then pitting corrosion initiates. The AlMn particle in the pits is the cathode phase, while the α-Mg phase is the anode. There are a large number of Mg^{2+} and OH^- ions in corrosion pits. When Mg^{2+} ions reach the maximum solubility, $Mg(OH)_2$ will be formed and pH value of the solution will stabilize around 10.5. Hereby, saturated Mg^{2+} ion leads to precipitation of $Mg(OH)_2$ at the bottom of the pit. And partial Mg^{2+} and OH^- ions may diffuse out of the pit with H_2 evolution, leading to the formation of $Mg(OH)_2$ precipitate outside the rim of the corrosion pit. This is the so-called occluded cell.

When an extruded Mg-1.21Li-1.12Ca-1Y alloy was immersed in Hank's solution for the first 15 min, pitting corrosion occurred at the interface of Mg_2Ca/α-Mg instead of $Mg_{24}Y_5$/α-Mg. Although the potential of the Mg_2Ca particle is higher than that of the α-Mg matrix phase, the dissolution rate of Mg_2Ca particles is higher than that of α-Mg matrix phase. After subsequent immersion of 1 h, corrosion product film formed around the pitting corrosion and filiform corrosion occurred. Pitting initiation in Mg_2Ca phase α-mg matrix forms corrosion products such as $MgCO_3$, $CaCO_3$, $LiCO_3$, $Mg_3(PO_4)_2$, $Ca_3(PO_4)_2$ and Li_3PO_4 that can make the corrosion product layer denser. During the electrochemical reaction, the surface of Mg_2Ca releases OH^- and H_2 and Mg^{2+} is produced by the dissolution of α-Mg. As the reaction proceeds, $Mg(OH)_2$ precipitates on the surface of Mg_2Ca phase. Mg^{2+} reacts further with HCO_3^- or CO_3^{2-} to form $MgCO_3$ precipitates.

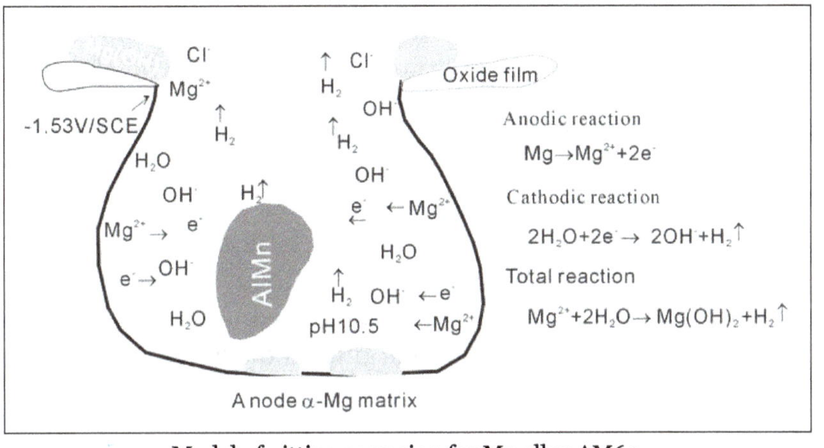

Model of pitting corrosion for Mg alloy AM60.

The pitting corrosion occurs on as-cast dual-phase Mg-Li-Ca alloy due to the presence of Mg_2Ca particles, whereas uniform corrosion appears after an extrusion processing. For the extruded Mg-Li-Ca alloy with refined microstructure, the tubular paths in the oxide film were jammed and sealed by the formation of an additional amorphous hydrated layer and the chemical compounds, that is, LiOH, $Mg(OH)_2$, $CaCO_3$ and $MgCO_3$, the PBR values of which are between 1 and 2. In this scenario, the formation rate of the corrosion products LiOH, $Mg(OH)_2$, $CaCO_3$ $MgCO_3$ and $CaxMgy(PO_4)z$ is greater than that of the degradation rate of the Mg-Li-Ca alloy, and hence the dissolution of the alloy may be readily self-inhibited.

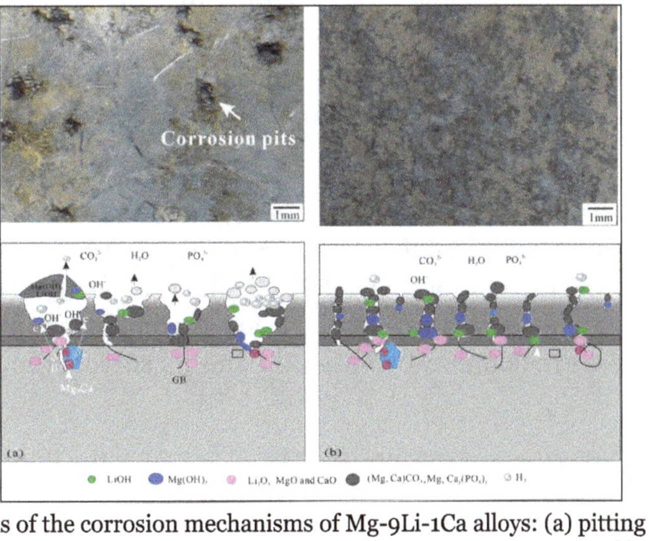

Schematic diagrams of the corrosion mechanisms of Mg-9Li-1Ca alloys: (a) pitting corrosion on the cast alloy and (b) tubular paths in the oxide film on the extruded alloy were jammed by corrosion products.

Pitting corrosion of Mg-Li alloys can be suppressed through grain refinement and formation of quasicrystalline or icosahedral phase (I-phase). I-phase eutectic pocket at GBs can suppress pitting corrosion and filiform corrosion of Mg-Li alloy with the presence of Zn and Y, regardless of Mg-6% Li-6% Zn-1.2% Y alloy containing a higher volume fraction of β-Li phase.

Filiform Corrosion

Filiform corrosion is also the most common type of corrosion for Mg alloys. Filiform corrosion of steel surfaces often occurs under coatings in high humidity environments. The filiform corrosion of Mg alloys is as common as pitting corrosion, but different from filiform corrosion of steels. Filiform corrosion is common on bare Mg alloy surfaces exposed to air and NaCl solution.

The pitting corrosion and filiform corrosion of Mg alloys usually occur simultaneously. Filiform corrosion originates from corrosion pits and extends forward along the active area. Corrosion of Mg alloys mainly is hydrogen evolution corrosion rather than oxygen absorption corrosion or oxygen concentration corrosion. Filiform corrosion is caused by active corrosive cells moving through the metal surface. Filiform corrosion of Mg is driven by the difference in oxygen concentration between filament head and filament tail, and thus proposed a filiform corrosion model for Mg. filiform corrosion occurs under protective coatings and anodic oxide layers, and pure Mg without coatings does not undergo filiform corrosion.

Chloride-induced fliform corrosion of organic-coated Mg is studied by Williams *et al.* The rate of filiform corrosion advance is shown to be insensitive to the presence of oxygen, but highly dependent upon the relative humidity of the holding environment. The filiform corrosion of Mg-Li alloy produces high concentration of OH^- reacting with Mg^{2+} and Li^+ to form $Mg(OH)_2$ and LiOH, which makes filament tail alkalized, passivates and covers the corrosion products of $Mg(OH)_2$. The hydrolysis reaction acidifies the corrosion pits and becomes a filament tip. Then the tip continues to propagate forward.

Intergranular Corrosion (IGC)

IGC does not occur on Mg and its alloys. It is regarded that the attack does not enter the grain along GBs, due to the fact that GBs are almost always cathodic relative to the grain interiors. Therefore, grains are anodic with regard to GBs, and the area next to GBs will be corroded. But this is not regarded as IGC, but granular corrosion, because corrosion tends to concentrate in areas adjacent to GBs.

However, intergranular stress corrosion cracking (ISCC) has been observed in the process of stress corrosion of Mg alloys, which indicates that IGC of Mg alloys is truly possible. It has been found that the morphology of IGC exists in as-extruded Mg alloy AZ80 after T5 aging treatment. The depth of this corrosion profile is shallow. The corrosion morphology is also due to the formation of micro-galvanic couple corrosion between $Mg_{17}Al_{12}$ and its surrounding matrix α-Mg, which is preferentially corroded.

From a microstructure perspective of the extrude Mg alloy ZK60, the fine second phase (Mg_2Zn) is dispersed along the GBs. Thus, pearl-like corrosion pits have been observed along the GBs in the extrusion direction. For zirconium-containing aluminum-free Mg alloy ZK60, the second-phase

Mg_2Zn on GBs is nobler relative to the α-Mg matrix phase, and corrosion occurs on the α-Mg matrix phase. Because zirconium is mainly distributed in the center of grains, the crystal is relatively stable. Further study discloses that the attack mainly occurs on the periphery of the grains, and no attack is seen to penetrate into the alloy along GBs. Therefore, the α-Mg matrix phase in the vicinity of GBs preferentially dissolves, which eventually leads to the whole grain peeling off. The cause of IGC is the formation of micro-galvanic corrosion between high potential finely distributed Mg_2Zn particles along the GBs and low potential matrix phase (α-Mg), which is typical IGC in Mg alloys for the first time.

Exfoliation Corrosion (EFC)

EFC is usually being thought of as a special IGC. EFC on wrought Al alloys has been extensively reported. Two essential factors exist in developing EFC: Elongated grains in microstructure and different potential between the matrix and intermetallic compound precipitates at GBs. However, EFC of Mg alloys has sparsely been reported. The EFC of cold-rolled Mg-14Li-1Al alloy appeared on both longitudinal and transverse surfaces. At the initial stage of corrosion, oxide film formed rapidly on the surface of magnesium alloy. Then the oxide film was attacked by chlorine ions, leading to the dissolution of Mg in local areas. There corrosion micro-cracks obviously emerge along GBs, which eventually results in denudation. After annealing, the corresponding EFC disappeared. The introduction of Al suppresses the exfoliation behavior of Mg-14Li alloy. The excessive addition of Al leads to a decrease in corrosion resistance. The optimal corrosion resistance could be achieved by adding 3 wt.% Al element.

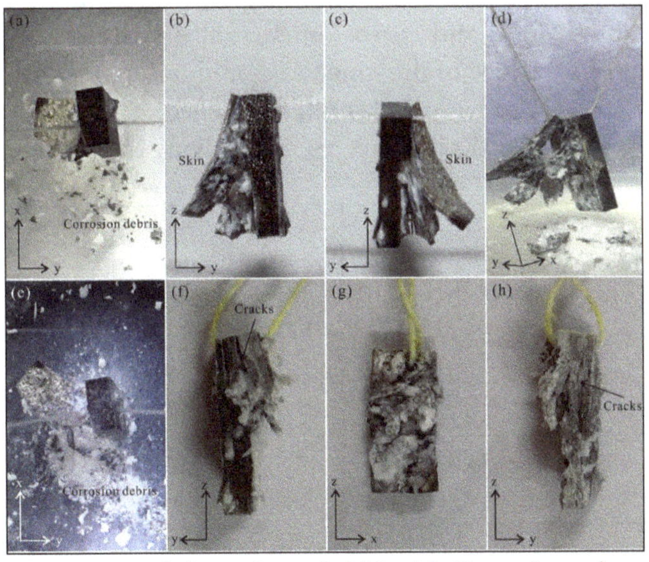

Corrosion macro-morphology of extruded Mg-1Li-1Ca specimens immersed in 3.5 wt.% NaCl aqueous solution after 90 days: (a)–(c); 120 days: (d) and (e); 150 days: (f)–(h).

Recently, the corrosion morphologies of an as-extruded Mg-1Li-1Ca alloy have been investigated after immersion in 3.5 wt.% NaCl aqueous solution for 90, 120 and 150 days. It is clearly that severe EFC occurs on the extruded Mg-1Li-1Ca plates and exhibits a "lifting-up" of the surface of the samples by the force of expanding corrosion products emerging at GBs just beneath the subsurface. A large amount of corrosion debris, mainly the carbonates, for example, Li_2CO_3, $CaCO_3$ and $Mg_5(CO_3)_4(OH)_2$, can be evidently seen between the layers and on the bottom of the container.

The optical images in figure disclose the bent outer layer and crack initiation sources and propagation directions. The cracks initiate along GBs or inside the grains, then coalescence, and arrest and propagation as well. The formation of voids or pores may be due to hydrogen blister and wedge effect resulted from the increase in volume of corrosion products.

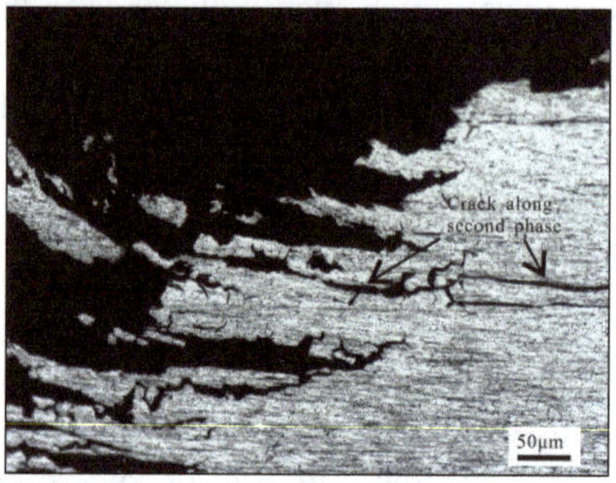

Optical image of cross-sectional morphologies of extruded Mg-1Li-1Ca alloy after an immersion in 3.5% NaCl aqueous solution for 21 days.

Figure demonstrates that open circuit potentials (OCPs) of the outer layer are much higher than those of the inner layer of as-extruded Mg-1Li-1Ca alloy. The gap in initial OCPs between them is greater than 88 mV/SCE. There is a risk of an increased corrosion rate for the less noble metal after galvanic coupling when the OCP difference is greater than 50 mV/SCE. Macro-galvanic corrosion may, thus, occur between the inner and outer layers. Figure shows the polarization curves of the inner zone and outer layer of the extruded Mg-1Li-1Ca alloy in 3.5 wt.% NaCl solution. Values of corrosion potential (E_{corr}) and corrosion current density (i_{corr}) were estimated through Tafel extrapolation fitting. E_{corr} values of the inner zone and outer layer are −1.639 ± 0.003 V/SCE and – 1.553 ± 0.009 V/SCE, respectively. Potential difference, ΔE_{corr} ($=E_{corr,\ outer\ layer} - E_{corr,\ inner\ zone}$ = 86 mV/SCE) of two layers is greater than 50 mV/SCE, implying the presence of galvanic corrosion.

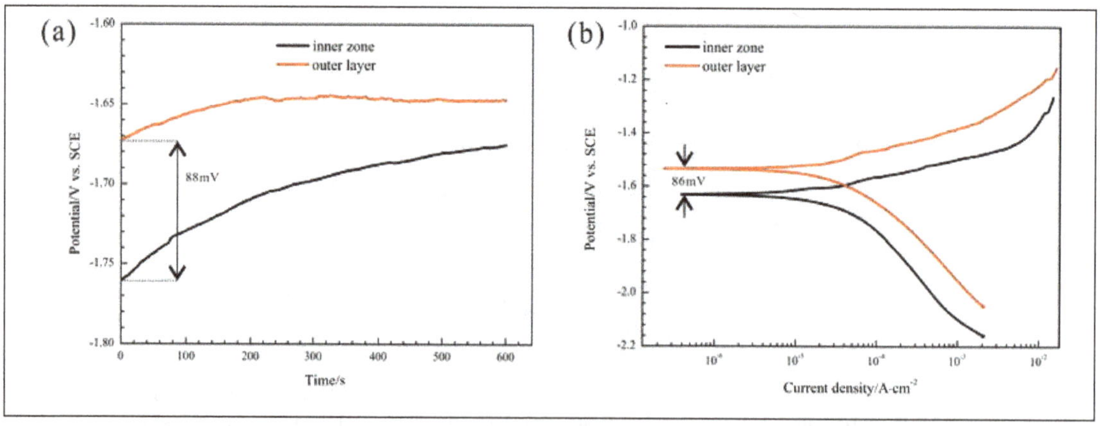

Open circuit potential as (a) function of immersion time and (b) polarization curves of inner zone and outer layer of extruded Mg-1Li-1Ca alloy in 3.5 wt.% NaCl solution.

KPFM was further used to evaluate Volta potential of the outer or skin layer and interior layer, as shown in figure. For the interior zone, the Volta potential of extruded Mg-1Li-1Ca alloy fluctuates over a narrow range, while the skin layer in purple color corresponds to the most negative Volta potential. The Volta potential difference implies the possible presence of macro-galvanic corrosion between the interior zone and skin layer. This result is consistent with that of electrochemical tests.

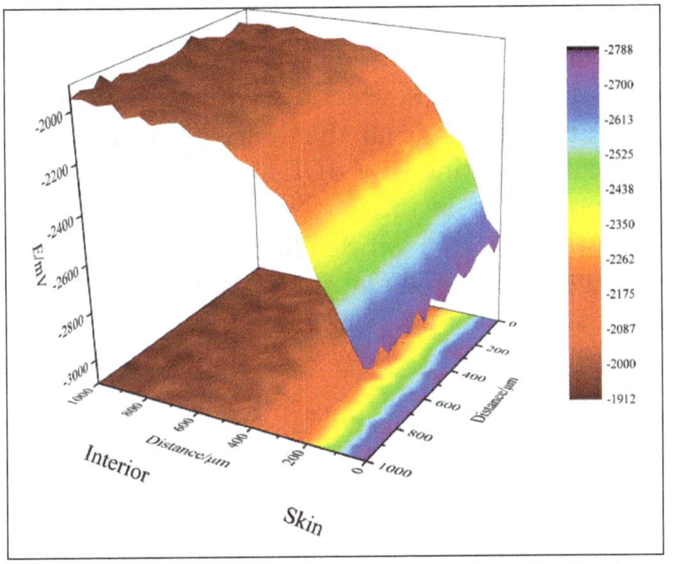

Surface Volta potential maps: Longitudinal section of extruded Mg-1Li-1Ca alloy, detected by KPFM.

The corrosion current density, i_{corr}, for the inner layer and the outer layer of the extruded Mg-1Li-1Ca alloy is 5.789 ± 0.598 × 10^{-5} A/cm^2 and 4.718 ± 0.259 × 10^{-5} A/cm^2, respectively. Microstructure of the alloy is characterized by a fibrous structure with elongated α-Mg grains and fine intermetallic compound Mg_2Ca. The skin layer with more refined grains shows better corrosion resistance, whereas the interior with coarse grains and Mg_2Ca particles distribute along GBs and extrusion direction in a line. The inner zone exhibits a negative OCP in comparison with the skin layer. When the inner zone and outer layer are coupled together, the inner zone acts as anode and is preferentially dissolved, whereas the skin layer plays a cathodic role and is protected.

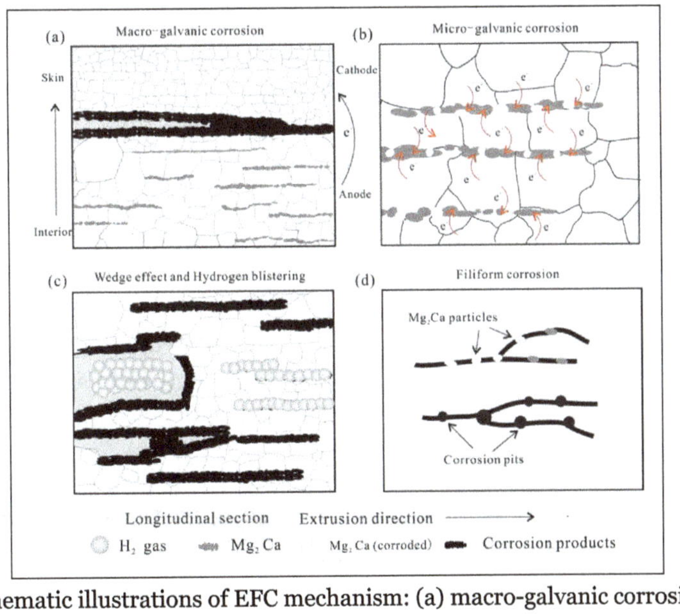

Schematic illustrations of EFC mechanism: (a) macro-galvanic corrosion between skin layer and interior zone, (b) micro-galvanic corrosion between Mg2Ca and α-Mg, (c) wedge effect and hydrogen blistering during EFC and (d) pitting corrosion and filiform corrosion.

Accordingly, several factors such as microstructure (i.e., elongated microstructures and second phases) and stress dominate EFC of extruded Mg-1Li-1Ca alloy. EFC mechanisms of the extruded Mg-Li-Ca alloy are schematically illustrated in figure. Exfoliation relates to both galvanic and wedge effects. On one hand, galvanic effect is resulted from the α-Mg grain and its neighboring second-phase Mg_2Ca particles, which are positioned along the extrusion direction at both GBs and the interiors. On the other hand, the wedge effect is caused by the enhancement in volume of corrosion products, consisting of $Mg(OH)_2$, $Ca(OH)_2$ and carbonates such as Li_2CO_3, $CaCO_3$ and $Mg_5(CO_3)_4(OH)_2$ piled up in the cracks. Exfoliation initiates at corrosion pits and propagates along corrosion filaments, in which filiform corrosion forms. Namely, EFC of the Mg-1Li-1Ca alloy emerged after pitting and filiform corrosion and was prone to stress corrosion. EFC of extruded Mg-Li-Ca alloys may be ascribed to a combining effect of pitting corrosion, filiform corrosion, IGC and SCC.

Crevice Corrosion

Occurrence of crevice corrosion of metals, such as Fe, stainless steel and Al alloys, is generally considered to be driven by oxygen concentration (oxygen reduction cathode reaction), while for Mg alloys, the change in oxygen concentration can be neglected. Therefore, crevice corrosion of Mg alloys has been neglected before. In 2011, crevice corrosion was firstly reported, which occurred in the crevice between specially treated pure Mg with unclear mechanism.

Stress Corrosion Cracking (SCC)

SCC of Mg alloys is the result of the combined action of electrochemistry and tensile mechanics. SCC can occur in any application when a stressed Mg-based component is subject to wet conditions. Basically, SCC mechanisms include anodic dissolution (AD) and hydrogen embrittlement (HE). The AD of Mg alloys is attributed to their high negative potential and the formation of a non-protective corrosion product film. HE is demonstrated to widely take place due to the fact that the existence of H is inevitable. Catastrophic failure is expected when hydrogen is introduced into Mg alloys, which decreases the SCC initiation stress to below the operating stress. So far, HE mechanism has not been clearly established. Pitting corrosion alleviated hydrogen into Mg alloy AZ80 and induced HE. It is suggested that the cracking mode of extruded Mg-7%Gd-5%Y-1%Nd-0.5%Zr (EW75) alloy is predominantly transgranular, but the partial intergranular stress corrosion cracking (IGSCC) can also be discerned at some localized area.

It is commonly recognized that transgranular SCC (TGSCC) of Mg alloys is a type of HE. HE models involve hydrogen-enhanced decohesion (HEDE), hydrogen-enhanced localized plasticity (HELP), adsorption-induced dislocation emission (AIDE) and delayed hydride cracking (DHC). The presence of hydride (MgH_2) phase on the fracture surface designates the DHC mechanism for a friction-stir welded magnesium alloy AZ31.

One of the results of plastic deformation of polycrystalline materials with hexagonal structure is the formation of many deformation twins. Although the deformation twin plays a very important role in other fracture processes, the function on stress corrosion process is ignored. For example, at room temperature, the FCP of polycrystalline Mg alloy is related to twin and GBs. Stress corrosion fracture surface of Mg-8.6Al alloy contains a lot of deformation twins. Twin rotation and hydrides occur during slow tensile stress corrosion of extruded Mg alloy AZ31. The relationship between SCC and deformation twin crystal plane is unclear.

Alloying elements have different impacts on SCC. Al is the most important element to produce stress corrosion sensitivity of Mg alloys. Mg-Al alloys have the highest stress corrosion sensitivity, and the sensitivity increases with Al content. An Al content of Mg alloys above 0.15–2.5 wt.% will result in SCC, with the maximum concentration at 6 wt.% Al. Zn also results in SCC sensitivity of Mg alloy. The addition of zirconium or RE elements (without Al) for Mg-Zn alloys such as ZK60 and ZE10 has moderate SCC resistance. Mg-Mn alloys are the most resistant to SCC due to the absence of Al and Zn.

Mg alloys also undergo SCC in distilled water. Some anions can also accelerate stress corrosion of Mg alloys. SCC sensitivity of Mg alloys in 0.1 M neutral salt solution decreases in the following order: $Na_2SO_4 > NaNO_3 > Na_2CO_3 > NaCl > CH_3COONa$.

The SCC behaviors of Mg alloy ZK60 under different conditions, that is, thin electrolyte layer (TEL) and bulk solution, had been investigated. Results indicated that Mg alloy ZK60 in the bulk solution exhibits a higher SCC susceptibility with a combined SCC mechanism of weaker AD and stronger HE compared to under TEL. They further revealed that cracking mode under both conditions was dominated by TGSCC feature and related to different HE mechanisms.

The effect of bovine serum albumin (BSA) to SCC of Mg alloy AZ91D in Hanks' solution was studied in. The alloy was susceptible to SCC with and without the presence of BSA. In the presence of BSA, the alloy suffered greater SCC on account of greater susceptibility to pit formation.

Corrosion Fatigue (CF)

CF is a key factor determining the service life of structural metals under cyclic mechanical loads. For example, car wheels must be resistant to CF. Air seems to be an aggressive atmosphere for Mg alloys. Microstructure and loading frequency have a critical influence on fatigue life and fatigue crack propagation rate (FCPR) of Mg alloys in ambient atmosphere. Since corrosion is time-dependent, CF of Mg alloys are sensitive to loading frequency. Under 10 Hz in air, fatigue life for as-extruded Mg alloy AM60 has a certain relationship with frequency. But when frequency is above 10 Hz, fatigue life is independent of frequency. This may be due to the slip or twinning of plastic deformation of Mg alloys at low frequencies and the formation of the intrusions or extrusions on the sample surface. This time seems to be enough to form an oxide film, which leads to an irreversible plastic deformation in a fatigue process. Because the plastic deformation speed is higher at a higher frequency, there's no sufficient time for extrusion of Mg alloys to react with oxygen in air to form an oxide film. Therefore, the formation of oxide film in a process of plastic deformation can be avoided at high frequency.

An investigation demonstrated the coalescence of micro-void FCP mechanism of the extruded AZ80 at a constant load amplitude fatigue test by virtue of scanning electron microscopy (SEM), transmission electron microscopy (TEM) and Auger electron spectroscopy (AES). FCPR increases with a decreasing loading frequency and an increasing load ratio. AES results designate that fatigue can markedly facilitate the fracture oxidation of magnesium alloys. In addition, the occurrence of crack closure is ascribed to plasticity instead of oxide film.

The pH value and Cl^- ion-containing solutions can significantly affect the fatigue life of Mg alloys. Fatigue life of as-extruded Mg alloy AM60 in a neutral NaCl solution was the shortest and the longest in an alkaline NaCl solution. CF strength of Mg alloys AZ91HP, AZ91-T6 and AM60 in 3.5% NaCl solution decreases significantly in comparison to that in air. In addition, Cl^-, Br^-, I^- and SO_4^{2-} ions accelerate the corrosion fatigue crack growth of Mg alloys.

Relative humidity (RH) and temperature in air have an important effect on FCP for magnesium alloys. Below 60 RH%, fatigue strength is no affected. At 90 RH% and higher, however, fatigue strength is significantly reduced. The effect of RH on the fatigue strength of extruded Mg alloy AZ61 has been investigated. At 55 RH%, the fatigue limitation is 145–150 MPa between 20 and 50 °C. At 80 RH% fatigue fracture occurs at the stress below the fatigue limitation, which means occurrence of CF. Corrosion pits are observed on the surface and fracture initiation area of the specimens. An increase in ambient temperature from 60 to 120 °C results in a higher FCPR.

Heat treatment (i.e., artificial aging T5) gives rise to an improvement in volume fraction of β phase, tensile strength and hardness of Mg alloy AZ80, and the fatigue life in aggressive solution at lower stress level.

For HCP metals, the combination of twinning and fatigue deformation with the existing twins is the main form of fatigue deformation. Dislocations and slip bands stack along the mechanical twin bands. Twins hinder the movement of dislocations, cause dislocation pile-ups, improve the internal stress of the matrix, and result in the combination of dislocation and form fatigue cracks eventually. Basically, the preferential initiation of fatigue cracks for cast Mg alloys develops at stress concentration sites such as the casting defects and pores. Fatigue cracks for wrought Mg alloys preferentially initiate at inclusions, cyclic slip bands and twin boundaries.

In air, fatigue crack of Mg alloy AZ80 initiates at the inclusions in the surface and subsurface. Corrosion pits are often the initiation sites for CF cracks. The formation of fatigue crack in an extruded Mg alloy AM60 relates to the AlMn particles. AlMn phase is the source of fatigue crack initiation in air. When it comes to a solution medium, the corrosion fatigue crack initiates at the corrosion pits around the AlMn phase particles. AD is the CF mechanism of extruded Mg alloys.

CF behavior of die-cast and extruded Mg alloys can be different. The fatigue life of die-cast and especially extruded Mg alloys AZ91D, AM50 and AZ31 is significantly decreased in 3.5% NaCl solution. Extruded alloys show a higher sensitivity to 3.5% NaCl solution compared to die-cast alloys. The CF life of extruded alloys is, however, longer than that of die-cast alloys.

Fatigue behavior of Mg alloys in simulated body fluids (SBF) is similar to that in 3.5 wt.% NaCl aqueous solution. Die-cast Mg alloy AZ91D showed a fatigue limit of 50 MPa at 10^7 cycles in air in comparison to 20 MPa at 10^6 cycles in SBF at 37°C. The extruded WE43 alloy possesses a fatigue limit of 110 MPa at 10^7 cycles in air compared to 40 MPa at 10^7 cycles in SBF at 37°C. The fatigue cracks initiate at the micro-pores in air and at corrosion pits in SBF, respectively. Corrosion resistance of such two alloys decreases under cyclic loading compared to that in the static immersion test. Demonstrated that sand-cast Mg alloy AZ91D, high-purity Mg, Mg-Ca and Mg-Zn-Ca alloys are susceptible to CF in SBF. And the fatigue cracks initiate at the microstructural defects such as inclusions, cast pores, micro-cracks and corrosion pits. HE is responsible for the cracking of Mg alloy AZ91D.

The influence of bovine serum albumin (BSA) on CF of Mg alloys has recently been reported. At stresses higher than the fatigue limit of Mg alloy AZ91D in air, mechanical factors play a predominant role. Hence, fatigue life is comparative for the alloys tested in air, Hank's solution in absence and presence of BSA. However, at lower stresses, Mg alloy AZ91D tested in Hank's solution + BSA discloses a longer fatigue life than that tested in a plain Hank's solution, indicating the inhibition impact of adsorbed BSA on corrosion and corrosion FCPR of the alloy.

Several strategies have been attempted to improve the resistance to CF. The shot-blasted Mg alloy AM60 exhibits an enhancement in fatigue strength in comparison to the die-cast one, which is ascribed to the fact that shot-blasting can induce hardening and compressive residual stress near surface region. Nevertheless, coatings such as calcium phosphate and MgF_2 film can provide little improvement in CF resistance. This is due to the degradation of the coating layer and existence of defects, which act as CF initiation sites under cyclical loading condition in SBF.

Erosion Corrosion

Mg alloys have low surface hardness and poor resistance to wear. Wear resistance relates to load amplitude and loading frequency. Wear loss of Mg alloys AZ91D and AM60B increases with normal loading force, but reduces with an increase in loading frequency. Corrosion and wear resistances can be greatly improved through micro-arc oxidation (MAO), a ceramic coating formed on the surface of Mg alloy AZ91. An $MgCO_3 \cdot 3H_2O$ film formed on Mg alloy AZ91 in HCO_3^- containing solutions is more compact and wear-resistant than porous $Mg(OH)_2$ film. the wear diameter of Mg alloy AZ91 and its MAO coating in HCO_3^- containing solution is smaller than that in NaCl solution, indicating that the formation of corrosion products in HCO_3^- bearing solution can resist micro-abrasion to a certain extent. That is, HCO_3^- ions can promote the wear resistance of Mg alloy AZ91 with and without MAO coating. There is, however, a critical concentration of HCO_3^- for enhancement in wear resistance. During wear corrosion, oxide particles peel off from MAO coating, which is detrimental to the corrosion resistance of MAO-coated Mg alloy AZ91. Ni-P coating especially added with inorganic nano-oxide particles (i.e., TiO_2 and SiO_2) can effectively improve the wear resistance of pure Mg.

Flow-induced shear stress (FISS) accelerates general corrosion and localized corrosion (including pitting corrosion and erosion corrosion) due to the enhanced mass transfer and mechanical force. FISS improves general corrosion rate, localized corrosion coverage ratios and depths and removal rate of corrosion products inside the corrosion pits compared to static conditions. Flow direction has a marked influence on corrosion resistance. More severe pitting corrosion and erosion corrosion were discerned on the back ends of the Mg-Zn-Ca alloy. And the corrosion product layer facing the flow direction was delaminated from Mg alloy AZ31 stent struts.

High Temperature Oxidation

Since Mg and its alloys are quite active and apt to react with surrounding oxidizing media such as oxygen, nitrogen and water vapor, oxidation-resistant and mechanical properties of Mg alloys are low at elevated temperatures. The oxidation resistance of metals depends mainly on the density and coverage of surface oxide films, which are also characterized by Pilling-Bedworth ratio (PBR).

PBR is related to the relative volume of oxide and its metal. When PBR is less than one, oxide film of the metal is porous and loose. When PBR is equal to or greater than one, a complete oxide film forms on the surface of the metal. Thus, the metal can be protected by such an oxide film. Basically, PBR of Mg oxide film is below one, which indicates that the oxide films of Mg alloys cannot protect substrates from the attack from the environment.

Alloying with Ca, Be and Sn as well as RE elements can improve the oxidation-resistance of Mg alloys significantly. Studies on the oxidation properties of Mg-Y-Sn, Mg-Y and Mg-Sn alloys at 500 °C show that Mg-Y-Sn and Mg-Y alloys demonstrate higher oxidation resistance compared with pure Mg and Mg alloy AZ31. The formation of a compact surface oxide film on Mg-Y-Sn and Mg-Y alloys after being oxidized at 500 °C for 6 h can effectively protect the alloys from further oxidation. Corrosion resistance of oxidized Mg-Y-Sn and Mg-Y alloys can be enhanced in comparison with the initial non-oxidized Mg alloys. Spark of micro-arc oxidation of Mg-Li-Ca alloys initiates at the α-Mg matrix with a lower content of Ca. This is due to the ignition inhibition of Ca.

There is, however, an issue on the oxide film of Ca-containing Mg alloys. Though the value of PBR for the oxides (MgO and CaO) of Mg-Ca alloys should be less than one, they exhibit good oxidation resistance. The working mechanism related to such an unexpected performance remains unclear.

Alternative Corrosion Modes

In many cases, several corrosion modes are observed on the same Mg alloys simultaneously or continuously. The corrosion process of Mg-Zn-Y-Zr alloy experiences three stages: galvanic corrosion, filiform corrosion and pitting corrosion in 0.1 M NaCl solution. The corrosion mechanism is governed by the second phase containing Mg, Zn and Y, which is much nobler than α-Mg matrix. There is a huge potential difference between the intermetallic compounds and their α-Mg matrix. At the first stage, the intermetallic compounds create galvanic corrosion. At the mid-stage, the intermetallic compounds prevent propagation of filiform corrosion. And at the last stage, the intermetallic compounds accelerate pitting corrosion. Different from the potential difference of Mg-Zn-Y-Zr alloy, the filaments of filiform corrosion on Mg-8Li alloy can extend a long distance due to a low potential difference between β-Li phase and α-Mg phase.

Similarly, different types of corrosion emerge alternatively for Mg-Gd-Zn alloys. Galvanic corrosion appears near second phase of $(Mg, Zn)_3Gd$ in the initial immersion, and filiform corrosion emerges after an immersion of 5 h. The difference of IGC developed in 16-h immersion is ascribed to the dissolution of $(Mg, Zn)_3Gd$ phase.

After a long-term immersion, the extruded Mg-Li-Ca alloys experience synergistic influence by pitting corrosion, filiform corrosion and EFC as well as SCC.

Corrosion morphology changes with the constituents of the films and volume fraction of the second phases. For Mg-7Sn alloy, the corrosion mode and rate are concerned with the quantity of the second phase Mg_2Sn and tin content of the α-Mg matrix. Pitting corrosion occurs when most of tin is in the presence of Mg_2Sn phase, while filiform corrosion emerges if most of tin is solutionized in the α-Mg matrix.

Application of Magnesium Alloys

Medical Applications

In the 1st half of last century, magnesium was first introduced in the medical industry as an orthopedic biomaterial. There are many characteristics and properties that make magnesium a very attractive option for use in implants and similar applications. Other common implant materials have densities that range from 3.1-9.2g/cm3, whereas the density of natural bone is 1.8-2.1g/cm3 (112.37-131.10lb/ft3). Magnesium alloys are much more comparable, at a density of 1.74-2.0g/cm3 (108.62-lb/ft3). Magnesium is also much more akin to natural bone than other materials in regards to fracture toughness, elastic modulus and compressive yield strength. Not only does magnesium provide the mechanical and physical properties desirable in these applications, it also exhibits some special characteristics.

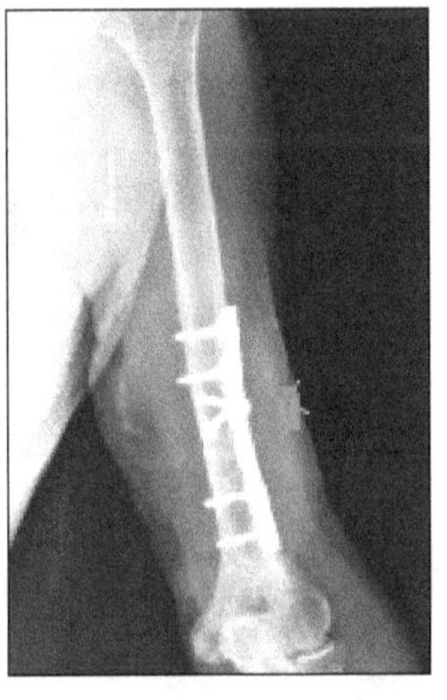

Magnesium is found naturally as an ion in the human body equating to about one mole in a 155lb (70kg) person, half of which is stored within bone tissue. Magnesium within the body assists in metabolic reactions, has good biocompatibility, and is nontoxic. In addition, uncoated magnesium implants can be biodegradable in bodily fluids through corrosion, which eliminates the need for a further surgery to remove implants. Application of protective coatings can prevent corrosion issues in situations where a more permanent solution is needed. Research and testing of different alloys and formulas for protective finishes is currently in progress with the aim of increasing the array of ways in which magnesium can be used in medical applications.

Electronic Applications

In the current market, small and portable electronics are trending upward. The demand for compact devices that can be easily transported is booming and magnesium is often a key component in meeting this demand. Many magnesium alloys being used to replace plastics are just as light, but they are much stronger and more durable. Magnesium is also better in regards to heat transfer and dissipation as well as its ability to shield electromagnetic and radio frequency interference. Many electronics require parts or casings with complex shapes which are possible with magnesium. Cameras, cell phones, laptops and portable media device housings are all common applications in addition to the arms of hard drives.

Sports Applications

Similar to its applications in other industries, magnesium is prized for use in sports equipment due to its light weight and impact resistance. Magnesium also has the ability to be formed into intricate shapes, which is ideal for use in golf clubs, tennis rackets and the handles of archery bows. The damping effects of the alloys also make it a good candidate for bicycle frames and the chassis of in-line skates, since the magnesium can absorb shock and vibration. This absorption allows cyclists to exert less energy and enjoy a smoother, more comfortable ride. Magnesium vaulting poles have also come into production as they have minimal twisting due to their high torsional strain resistance.

Transport Applications

Magnesium alloys have always been attractive to designers due to their low density, only two thirds that of aluminum alloys in the aerospace industry and therefore can be an innovation technology if used for low weight airframe structures. This has been a major factor in the widespread use of magnesium alloy castings and wrought products.

New light materials are currently inserted in world strategies of transport vehicle industry since the environment necessities for pollution and reduction of fuel consumption. Therefore the industry takes part of the risk of development of such alloys but, in fact, some of this has been made at academic level. Some aspects of the necessities and characteristics concerning those alloys are: low costs, insulation (sound and thermal), impact safety, deformation strength, recyclability and guarantee (to aging as example).

However to use this low weight material several mechanical properties have to be increased and the technological behavior improved. A further requirement in recent years has been for superior corrosion performance and dramatic improvements have been demonstrated for new magnesium alloys. Improvements in mechanical properties and corrosion resistance have led to greater interest in magnesium alloys for aerospace and specialty applications.

Magnesium alloys are also used in many other engineering applications where having light weight is a significant advantage. Magnesium-zirconium-aluminum alloys tend to be used in relatively low volume applications where they are processed by sand or investment casting, or wrought products by extrusion or forging.

Actually it's well known that there are a growth in primary production of magnesium due to a supplementary application of your products in market. A total 2010 production of primary magnesium in the world was approximately 809,000 tons. The main producer are China 654,000 tons, USA 45,000 tons, Russia 40,000 tons, Israel 30,000 tons, Kazakhstan 20,000 tons and Brazil 16,000 tons.

The automobile have more than a hundred years since its invention and the light alloys have been utilized since 1915. In the 30's the aluminum did substitute almost completely the casting iron as main component of pistons. In the same decade, in Germany, magnesium alloys were utilized in the production of camshaft and in the gear box, which leads, at that time, the total decrease of weight at least 7% of the total of the automobile.

Magnesium alloy developments have traditionally been driven by aerospace industry requirements for lightweight materials to operate under increasingly demanding conditions. In the last decade several heavy magnesium parts have been assembled in passenger cars, such as gear box housings and crank cases.

Considerable research is still needed on magnesium processing, alloy development, joining, surface treatment, corrosion resistance, and mechanical properties improvement. Surface coatings produced for magnesium die-casting by hexavalent chromium baths have been used to provide stand-alone protection and as a pretreatment for painting. Teflon resin coating has been developed for Mg alloys; initially the coating is obtained with an aluminum vapor deposition and finish treatment with a Teflon resin coating. The Teflon resin coating is a low cost, chromium-free corrosion resistant coating for magnesium alloys. The coating not only has corrosion resistant properties, but also good lubricity, high frictional-resistance and non-wetting properties.

These new projects are concerned on the modification of existing and the development of new magnesium wrought products (sheets and extrusions), that provide significantly improved static and fatigue strength properties for lightweight fuselage applications (Development of material models and failure criteria for the prediction of forming processes, plastic deformation and failure behavior of components; material adapted design and the evaluation of structural behavior to close the process and development chain for aeronautic components).

The specific strength properties of these innovative materials are required to be higher than

AA2024 for structural applications (secondary structure) and higher than AA5083 for non-structural applications. At the beginning of the project new alloys will be developed and experimental alloys will be tested. Appropriate manufacturing (rolling, extrusion), forming and joining technologies require development, simulation and validation for the innovative material and application. The technological objective is a weight reduction of the fuselage structure, system and interior components up to 30 %. The strategic objectives are an increase in the operational capacity of 10 %, a reduction in the direct operating cost of 10 % and finally a reduction in the fuel consumption of 10 % and therefore a reduced environmental impact with regard to emissions and noise.

Regarding corrosion is also a problem to be solved with newly adapted surface protection systems according to aerospace requirements and advanced design concepts. The addition of chemical elements and special surface treatments in magnesium to avoid flammability (auto-ignition temperature); Mg alloys with eutectic phases, low melting temperature, ignite first with comparative investigations to aluminum alloys.

For those purposes there are nowadays some new techniques for the production of this kind of light alloys, notably: powder metallurgy (PM), which is essentially the process that leads small particles (from sizes in a range from nanometers to tens of micrometers) to a consolidation process that results in a uniform and fine grained structure material; spray forming techniques that relies in production processes which allows the production of parts and components by deposition of sprayed particles in semi-molten state that will consolidate in an injection mold with a substrate and metal foam. In addition the nanostructured materials have been the utmost in materials development due mainly to their particular relationship between microstructure of grains or hyperfine particles and mechanical properties, differentiated from those produced by conventional processes.

The development of solid state powder metallurgy makes it possible to increase the values of solute above the limits of conventional metallurgy arising a new era for the magnesium and aluminum alloys. Increasing the amount of "thirds" alloying elements gives to these alloys new characteristics such as high deformation rates at room temperature, refinement of microstructure (hyperfine grains or nanostructured materials), new conditions for cold and hot work processing, welding, wear resistance, increase of hardness as examples.

Also, recyclability of magnesium and aluminum alloys is a growing factor in choice for utilization in automotive industry, since nowadays it's possible to recycle up to 99% of parts and pieces made of magnesium and aluminum which reduces the costs with transportation and this is also good for the environment.

News production technologies search for minimize the quantities of energy employed in its processes and the recyclability shows an economy of 95% of energy comparing to that produce primary magnesium and aluminum generating an "ecological compensation".

Another influence in research for new technologies of materials and light alloys relies on the shortening of life of models. New materials and alloys are usually more expensive than commercial materials, so there is direct needs of investment in reduce the costs of production and development of such alloys allowing the utilization in large scale in the automobile production processes.

Despite the fact that the introduction of light alloys and new technology light alloys is a tendency not changeable, the utilization effective today still is more applicable to competition or sportive cars and motorcycles, due to the high costs. However if compared along the existence of automobile the employment of light alloys rise exponentially from the earliest up to the latest commercial model. The influence of 1970's in the development of such technologies is notable comparing with the few kilograms used in the first automobiles. All those factors contribute to new researches and development of this class of materials for structural and mechanical applications in automotive industry (transport).

Global effort in magnesium materials development has diversified in recent years, targeting new automotive, aerospace, and also in biomedical applications. The 1990s saw a strenuous effort in the tissue to regenerate without the use of a second surgery to remove the repair implant. Research in this field is focusing on the development of new magnesium alloys with low rates of bio-degradation. Emerging areas of application for magnesium are the input for global effort in magnesium materials development. Interdisciplinary and international collaborations will be the key in developing magnesium to its full potential as an environmentally friendly structural material with wide-ranging applications.

A very interesting program to benefit the magnesium utilization is MagForming, an international project (12 different partners) with the aim to advance the state of the art in forming methods for a range magnesium alloys in extruded, sheet and plate forms using aerospace prototypes. This includes forging, superplastic forming, roll bending, pad forming, deep drawing and creep forming. The project aims to develop best practices for all of these methods via production of aerospace prototype parts. The data results in production of useful parts for testing, potentially leading to the use of magnesium in the system and structure areas from which the parts have been taken.

Magnesium alloys had been widely used in aircrafts up to the 1950s. Due to some major drawbacks such as high corrosion sensitivity and, under extreme conditions, flammability other materials, both metals and plastics, have increasingly replaced magnesium. Airbus is now reappraising the introduction of magnesium alloys in aircraft. Additionally Airbus is developing and investigating approaches for a lab scale fire testing procedure with the scope to ensure compliance with FAA-regulations on aircraft fire safety.

Comparative and Effective use of New Mg Alloys

New light materials are effectively nowadays inserted in world strategies of automotive industry since the environment necessities for pollution and reduction of fuel consumption. Therefore the industry takes part of the risk of development of such alloys but, in fact, some of this has been made at academic level. It's possible to enumerate some aspects of the necessities and characteristics concerning those alloys: low costs, insulation (sound and thermal), impact safety, deformation strength, recyclability and guaranty (to aging as example). All those aspects are linked with the increasing of new automobile models and reflect in production programs more and more complexes.

Different joining techniques were applied to magnesium wrought semi-finished products, in order to promote their introduction on aeronautical structures. Airbus has performed some first tests to join magnesium sheets by friction stir welding. In general the alloy AZ31B (Mg-3.0%Al-0.3%Mn) is quite easily weldable by different processes. Using laser beam welding an AZ61 (Mg-5.9%Al-0.5%Zn-0.2%Mn) filler wire is advantageous for the mechanical properties to weld this alloy.

Another influence in research for new technologies of materials and light alloys relies on the shortening of models life. New materials and alloys are usually more expensive than commercial materials, so there is direct needs of investment in reduce the costs of production and development of such alloys allowing the utilization in large scale in the automobile production processes.

The increase in the potential application of magnesium profiles is strongly dependent on the question of whether established forming processes for aluminum and steel can be changed to magnesium and its alloys. Broad-spectrum applications of magnesium alloys in the automotive industry are casting products.

Despite the fact that the introduction of light alloys and new technology light alloys is a tendency not changeable, the utilization effective today still is more applicable to competition or sportive cars and motorcycles, due to the high costs previously mentioned. However if compared along the existence of automobile the employment of light alloys rise exponentially from the earliest up to the latest commercial model. The influence of 1970's in the development of such technologies is notable comparing with the few kilograms used in the first automobiles.

Recently the weightiness of light alloys, for example, in an automobile is near 90 kg in Europe, 120 Kg in United States and 42 kg in Brazil, but increasing year-to-year. Nearly 90% or more from the weight relies on aluminum alloys, but there is a rapidly increase in the magnesium and a slightly in titanium alloys in the total amount used. All those factors contribute to new researches and development of this class of materials for structural and mechanical applications in automotive industry.

Traditionally the main usage for magnesium and magnesium alloys has been for aluminum alloying, high pressure die casting and steel desulphurization. Over the last 10 years the demand for magnesium and its alloys has grown at an average rate close to 5% per year. The die casting industry which expanded at a rate of over 10% per year was mainly responsible for this steady growth of the whole industry. This remarkable growth was possible because of the stable and relatively inexpensive supply of magnesium from China. This low costs supply has changed at the end of 2007 and early 2008. During that period the base price of magnesium has tripled. Special attention will be given to the growth potential of magnesium alloys for components which will be driven most likely by environmental regulations from governments.

As the lightest structural materials, magnesium alloys are well suited for the car industry and also good fuel economy is essential. The selection of a new alloy for a vehicle component should be based on technical requirements and targeted cost. In reality, this selection process is complicated and depends very much on the relative weight given to a specific property, which is part of the combined desired properties and final targeted cost. This task becomes even more complicated if alternative material systems such as aluminum alloys are considered for the same applications.

Several new magnesium alloys have been developed recently for high temperature applications to obtain an optimal combination of die castability, creep resistance, mechanical properties, corrosion performance, and affordability. Most of the new alloys can only partially meet the required performance and cost. The ZE41 alloy (gravity-casting applications) has moderate strength and creep resistance combined with good castability. Although this alloy exhibits poor corrosion resistance, it is still preferred for certain applications.

Although the most commonly used magnesium die-casting alloys are of the AZ and AM series, improved elevated-temperature performance is required (gearbox housing, intake manifolds, oil pans, transfer cases, crankcases, oil pump housing). Insufficient creep strength of alloys can causes poor bearing-housing contact, leading to oil leaks and increased noise and vibration.

The use of magnesium alloy casting in the automobile industry expands at an impressive rate in this decade, which can manage with the energy and environment problems. Alloy AZ91 (Mg-9Al-0.8Zn-0.2Mn) is the most favored magnesium alloy, being used in approximately 90% of all magnesium cast products.

There are two patented magnesium alloys: Mg-Al-Ca-Sr based alloy (MRI 153M) and Mg-Al-Ca-Sr-Sn based alloy (MRI 230D). The MRI 153M is a beryllium-free, creep-resistant alloy capable of long operation at temperatures up to 150 °C under high stresses (substantially superior to those of commercial alloys). The MRI 230D is a die-casting alloy developed for use in automotive engine blocks operating at temperatures up to 190 °C. The alloy has excellent creep resistance combined with good castability, high strength, and superior corrosion behavior. The results obtained show that MRI 230D and A380 exhibit similar tensile creep behavior at 150–175 °C under stress of 70 MPa.

Concerning the whole aeronautic industry, due to the fact that weight reduction is a very important objective for strengthening the competitiveness, several alternatives to obtain weight reduction has to be investigated (welded or bonded airframes; use of metal laminates; structural plastics; fiber reinforced composites).

The non-metallic materials application in selected areas is not conceivable due to restricted properties under low or elevated temperatures, missing electrical conductivity or low damage acceptance. Fiber reinforced plastics are a relatively lavish material only used for primary structure applications with highest demands.

The family of magnesium alloys and especially magnesium wrought materials can be an excellent alternative because of their low density, good mechanical properties, moderate cost and metallic character (in respect of manufacturing, repair, maintenance compared to composites).

In the past decade a lot of research activities and development projects have been carried out working on magnesium cast materials mainly for transport applications. There were only very few activities on magnesium wrought products like sheets, extrusions or forged parts. The alloy spectrum of magnesium wrought alloys is still very restricted.

Aeronautic requirements and applications of wrought products have been evaluated only in a few projects. Increasing the research on magnesium wrought alloys will promote a new class of metallic materials for aeronautical applications to win the competition against plastics and fiber reinforced plastics. Therefore, the variety of offered metallic materials will be enlarged, not only for aircrafts, but also for space, military and satellites applications.

To reach this objective magnesium has to deliver meaningfully higher weight specific mechanical properties compared to aluminum. The aims for aluminum replacement can be divided into two different steps in respect of time scale and risk.

A replacement of medium strength 5XXX aluminum alloys for cockpit and cabin applications and another possible replacement of medium to high strength 2XXX aluminum alloys for secondary structure or non-pressurized fuselage applications.

Forming and joining technologies require development, simulation and validation for the innovative material and technologies commonly used within aeronautic industry. Recently Hombergsmeier presented the requirements of new alloys concerning property temperature systems and structural applications.

Table: Requirements for Mg alloys for transport applications.

Properties	T(K)	Materials	
		Structural applications	Systems applications
Tensile Ultimate Strength	RT	450 Mpa	275 - 350 MPa
Tensile Yield Strength	RT	350 Mpa	200 - 300 MPa
Elongation to fracture	RT	16 - 18 %	12 - 16 %
Yield Strength	423	0.9 YTS	0.9 YTS
Compressive Yield Strength	RT	1.1 YTS to 0.9 YTS	1.1 YTS to 0.9 YTS
Failure under compression	RT	alike: Al 2024 T3	alike: Al 5083
Specific Weight	RT	1.75	1.75
Residual Strength	RT	alike to 2024 T3	n. a.
Fatigue Crack Growth	RT	alike to 2024 T3	n. a.
Fatigue Limit (K_t=1.0, R=0.1)	RT	140 Mpa	160 MPa

The most promising new alloy systems selected due to corrosion behavior, green friendliness and mechanical performance for further investigation as wrought products were Mg-Al-Zn, Mg-Zn-Zr-Re and Mg-Y-Re. An undertaking processing for magnesium alloys is the application of severe plastic deformation (SPD) that makes available a useful tool for introducing very significant grain refinement in bulk metallic materials. Even though several SPD techniques are now accessible, processing by equal-channel angular pressing (ECAP) is especially attractive because it is easily scaled-up for use with large samples. Since grain refinement is generally believed to have beneficial effect on properties of Mg alloys, fabrication of bulk ultrafine-grained (grain size less than 1μm) alloys using the new conceptual metal forming process (SPD) should attract considerable attention.

Magnesium Alloys for Cycle Applications

Bicycle frames have gone from wood to steel to advanced alloys to composites in the last century and a half. Bikes themselves have gone from simple curiosities to serious modes of transportation in developing countries and a major form of sports and recreation in developed ones. The improvements in automotive and aerospace industry are now helping to increase a rebirth in the bicycle industry. Similarly, mass production techniques derived for the automotive industry have helped bring the cost down but, many of the original materials and production processes used today remain unchanged from over a hundred years ago.

The basic properties of magnesium alloys propose that they would keenly find use in bicycles. Magnesium alloys have low density and a high strength to-weight ratio, are readily extrudable, and some alloys are highly weldable. A lack of information about wrought magnesium alloys, a lack of suppliers to complete the production process, and a lack of information in how to set up a mechanized plant to process wrought magnesium alloys have contributed to only limited market dissemination by magnesium into bicycle applications.

Many doubts limit the recognition of welded magnesium alloy tubes as a substitute bicycle frame material. Most bicycle industry believes that magnesium is not weldable and is very brittle. But some alloys are weldable and it has been recognized that the ductilities of common magnesium alloys are equivalent to aluminum alloys, for example, 6061 and 7075 aluminum alloys.

A common misunderstanding is related to wellbeing, with a conviction that contact with oxygen or water will cause explosions. Also that magnesium alloys lack sufficient stiffness to make a bicycle frame. Actually start the acceptance larger diameters to compensate lower stiffness and tensile strength. Concerning the difference between the Young modulus of aluminum and magnesium is much less than the difference between steel and aluminum. With applied basic design principles, it is assumed that the general requirement for product improvements will press on bicycle manufacturers to more seriously consider wrought magnesium as an attractive alternative to the utilized common materials (steel, aluminum, carbon fiber composites).

Magnesium frames may be manufactured by die casting or by welding extruded and welded tubes. Die-casting requires the recovery of die costs that is expensive, but cast magnesium can offer economy with large production volumes. So a manufacturing route could offer a low-cost transportation solution for a developing country.

The wrought magnesium alloys generally offer improved elongation and superior forms. Wrought tubes formed and bent into frame tubes represent a highly flexible and economical approach. Also the weldability rates of magnesium alloys are generally higher than competing aluminum or titanium alloys, or even alloy steels (Welding is the primary method of joining planned for magnesium alloys. His efficiency is very high and many Mg alloys not require post heat treatment).

Among the high damping properties and low density of magnesium it can easily exceed any of the current metals in ride quality leading to an improved fatigue life. Extruded profile designs need to reduce weight and maintain dent resistance in specific key functional areas. Magnesium alloys have a range of potential benefits to offer the bicycle designer, ranging from low density, high specific strength, and dent resistance to extrudability, machinability, and weldability. Finally, making bicycles that win races is the key to providing magnesium a lasting foothold in the cycle industry.

In other areas of the bicycle, rims have been made using Elektron ZM21 and AZ61. Pedals and handlebar stems have also been developed. Wheel hubs, seat posts, brakes, cranks, and ven forks have been produced by companies that specialize in industrialization of high-quality products. Clearly, pioneering an entire class of metal was not well suited to replacement strategy or ad-hoc engineering. Most of the initial efforts have not been successful because rushed planning combined with limited infrastructure ultimately led to problems.

Other Applications

Magnesium is utilized in optical and hand-held tool design. Wearers of spectacles often desire an aesthetically pleasing frame that is not felt during day-to-day wear but is sturdy enough to handle being removed and replaced throughout the day. Rifle scopes and binoculars must be stable in order to be effective and a light weight is also beneficial in their marketability. Magnesium allows for these design criteria to be met.

Hand-held work tools and devices such chain saws, hand shears, hand drills, pneumatic nail guns and weed whackers are all perfect candidates for magnesium applications. The low density, comparative strength and vibration damping capabilities are all desirable characteristics in creation of handheld tools such as these.

Magnesium has also found its way into small household appliances such as vacuums. By utilizing this material for a vacuum head, it will be light enough to easily move around as well as being dent resistant in normal use when coming in contact with furniture and other obstacles.

As the inherent benefits of magnesium applications are further realized, the ways in which it can and will be used become numerous. Demand for sustainable, lightweight and recyclable materials is ever increasing with the push towards environmentally-conscious products, which can only be beneficial for the magnesium industry.

References

- Metal-profile-magnesium-2340142: thebalance.com, Retrieved 06 May, 2019
- Magnesium-Notable-characteristics-of-element-and-compounds: newworldencyclopedia.org, Retrieved 23 June, 2019
- Microstructure-and-properties-of-casting-magnesium-alloys-designed-to-work-in-elevated-temperature, magnesium-alloys-selected-issue: intechopen.com, Retrieved 12 April, 2019
- Corrosion-types-of-magnesium-alloys, magnesium-alloys-selected-issue: intechopen.com, Retrieved 16 February, 2019
- New-features-on-magnesium-alloys-application-of-magnesium-alloys-in-transport: intechopen.com, Retrieved 14 March, 2019

5

Titanium and Beryllium Alloys

Titanium is a lustrous, silver colored transition metal which has a low density and high strength. Beryllium is an alkaline earth metal which is strong, brittle and lightweight. It is used to form alloys along with nickel, iron, copper and other intermetallic compounds. The topics elaborated in this chapter will help in gaining a better perspective about the key properties and diverse uses of titanium and beryllium alloys.

Titanium

Titanium (chemical symbol Ti, atomic number 22) is a silvery-white, lustrous metal. Light and strong, it is resistant to corrosion, including corrosion by seawater and chlorine. It is used in strong, lightweight alloys (most notably with iron and aluminum), and in powdered form to other materials, such as graphite composites. Its most common compound, titanium dioxide, is used in white pigments. Examples in which white pigment, consisting of titanium oxide, is used, are correction fluid and commonly used white paint to repaint walls. It is also used in toothpaste, white road marking paints, and white fireworks. Substances containing titanium are called titaniferous.

The element occurs in numerous minerals with the main sources being rutile and ilmenite, which are widely distributed over the Earth. There are two allotropic forms and five naturally occurring isotopes of this element—^{46}Ti through ^{50}Ti with ^{48}Ti being the most abundant (73.8 percent). One of titanium's most notable characteristics is that it is as strong as steel but is only 60 percent its density. Titanium's properties are chemically and physically similar to those of zirconium.

Occurrence

Titanium metal is always bonded to other elements in nature. It is the ninth-most abundant element in the Earth's crust (0.63 percent by mass) and is present in most igneous rocks and in sediments derived from them (as well as in living things and natural bodies of water). It is widely distributed and occurs primarily in the minerals anatase, brookite, ilmenite, perovskite, rutile, titanite (sphene), as well in many iron ores. Of these minerals, only ilmenite and rutile have significant economic importance, yet even they are difficult to find in high concentrations. Because it reacts easily with oxygen and carbon at high temperatures, it is difficult to prepare pure titanium metal, crystals, or powder. Significant titanium ore deposits exist in Australia, New Zealand, Scandinavia, North America, and Malaysia. Large quantities have also been detected in the Kwale region in Kenya, deposits to which a Canadian firm, Tiomin, has mining rights.

Producer	Thousands of tons	percent of total
Australia	1291.0	30.6
South Africa	850.0	20.1

Canada	767.0	18.2
Norway	382.9	9.1
Ukraine	357.0	8.5
Total: top 5	3647.9	86.5
Total world	4221.0	100.0

This metal is found in meteorites and has been detected in the sun and in M-type stars. Rocks brought back from the moon during the Apollo 17 mission are composed of 12.1% TiO_2. Titanium is also found in coal ash, plants, and even the human body (while harmless, it is not believed to be an essential element).

Isolation

Because the metal reacts with air at high temperatures it cannot be produced by reduction of its dioxide. Titanium metal is, therefore, produced commercially by the Kroll process, a complex and expensive batch process developed in 1946, by William Justin Kroll. In the Kroll process, the oxide is first converted to chloride through carbochlorination, whereby chlorine gas is passed over red-hot rutile or ilmenite in the presence of carbon to make $TiCl_4$. This is condensed and purified by fractional distillation and then reduced with 800 °C molten magnesium in an argon atmosphere.

Titanium (mineral con4centrate).

A newer process, the FFC Cambridge Process, may replace the older Kroll process. This method uses the feedstock titanium dioxide powder (which is a refined form of rutile) to make the end product which is either a powder or sponge. If mixed oxide powders are used, the product is an alloy at a much lower cost than the conventional multi-step melting process. It is hoped that the FFC Cambridge Process will render titanium a less rare and expensive material for the aerospace industry and the luxury goods market, and will be seen in many products currently manufactured using aluminium and specialist grades of steel.

Titanium was purified to ultra high purity in small quantities when Anton Eduard van Arkel and Jan Hendrik de Boer discovered the iodide, or crystal bar, process in 1925, by reacting with iodine and decomposing the formed vapors over a hot filament to pure metal.

Titanium oxide is produced commercially by grinding its mineral ore and mixing it with potassium carbonate and aqueous hydrofluoric acid. This yields potassium fluorotitanate (K_2TiF_6) which is extracted with hot water and decomposed with ammonia, producing an ammoniacal hydrated oxide. This, in turn, is ignited in a platinum vessel, which creates pure titanium dioxide.

Common titanium alloys are made by reduction. For example; cuprotitanium (rutile with copper added is reduced), ferrocarbon titanium (ilmenite reduced with coke in an electric furnace), and manganotitanium (rutile with manganese or manganese oxides) are reduced.

Physical properties	
Phase	solid
Density (near r.t.)	4.506 g/cm^3
Liquid density at m.p.	4.11 g/cm^3
Melting point	1941 K (1668 °C, 3034 °F)
Boiling point	3560 K (3287 °C, 5949 °F)
Heat of fusion	14.15 kJ/mol
Heat of vaporization	425 kJ/mol
Heat capacity	(25 °C) 25.060 J/(mol·K)

Vapor pressure						
P/Pa	1	10	100	1 k	10 k	100 k
at T/K	1982	2171	(2403)	2692	3064	3558

Notable Characteristics

Titanium is classified as a transition metal. It is well-known for its excellent resistance to corrosion; it is almost as resistant as platinum, being able to withstand attack by acids, moist chlorine gas, and by common salt solutions. Pure titanium is not soluble in water but is soluble in concentrated acids. A metallic element, it is also well-known for its high strength-to-weight ratio. It is a light, strong metal with low density that, when pure, is quite ductile (especially in an oxygen-free environment), lustrous, and metallic-white in color. The relatively high melting point of this element makes it useful as a refractory metal. Commercially pure grades of titanium have an ultimate tensile strength equal to that of high strength low alloy steels, but are 43 percent lighter. Titanium is 60 percent heavier than aluminum, but more than twice as strong as 6061-T6 aluminum alloy; these numbers can vary quite substantially due to different alloy compositions and processing variables. It is fairly hard (although by no means as hard as some grades of heat-treated steel) and has a tendency to blunt cutting tools. Like those made from steel, titanium structures have a fatigue limit which guarantees longevity in some applications.

This metal forms a passive and protective oxide coating (leading to corrosion-resistance) when exposed to elevated temperatures in air, but at room temperatures it resists tarnishing. The metal, which burns when heated in air 610 °C or higher (forming titanium dioxide) is also one of the few elements that burns in pure nitrogen gas (it burns at 800 °C and forms titanium nitride). Titanium is resistant to dilute sulfuric and hydrochloric acid, along with chlorine gas, chloride solutions, and most organic acids. It is paramagnetic (weakly attracted to magnets) and has a very high electrical conductivity and thermal conductivity.

Experiments have shown that natural titanium becomes radioactive after it is bombarded with deuterons, emitting mainly positrons and hard gamma rays. The metal is a dimorphic allotrope with the hexagonal alpha form changing into the cubic beta form very slowly at around 880 °C. When it is red hot the metal combines with oxygen, and when it reaches 550 °C it combines with chlorine. It also reacts with the other halogens and absorbs hydrogen.

Atomic properties	
Crystal structure	hexagonal
Oxidation states	2, 3, 4 (amphoteric oxide)
Electronegativity	1.54 (Pauling scale)
Ionization energies (more)	1st: 658.8 kJ/mol
	2nd: 1309.8 kJ/mol
	3rd: 2652.5 kJ/mol
Atomic radius	140 pm
Atomic radius (calc.)	176 pm
Covalent radius	136 pm

Isotopes

Naturally occurring titanium is composed of 5 stable isotopes; ^{46}Ti, ^{47}Ti, ^{48}Ti, ^{49}Ti and ^{50}Ti with ^{48}Ti being the most abundant (73.8 percent natural abundance). Eleven radioisotopes have been characterized, with the most stable being ^{44}Ti with a half-life of 63 years, ^{45}Ti with a half-life of 184.8 minutes, ^{51}Ti with a half-life of 5.76 minutes, and ^{52}Ti with a half-life of 1.7 minutes. All of the remaining radioactive isotopes have half-lifes that are less than 33 seconds and the majority of these have half-lifes that are less than half a second.

The isotopes of titanium range in atomic weight from 39.99 amu (^{40}Ti) to 57.966 amu (^{58}Ti). The primary decay mode before the most abundant stable isotope, ^{48}Ti, is electron capture and the primary mode after is beta emission. The primary decay products before ^{48}Ti are element 21 (scandium) isotopes and the primary products after are element 23 (vanadium) isotopes.

Compounds

The +4 oxidation state dominates in titanium chemistry, but compounds in the +3 oxidation state are also common. Because of this high oxidation state, many titanium compounds have a high degree of covalent bonding.

Although titanium metal is relatively uncommon, due to the cost of extraction, titanium dioxide (also called titanium(IV), titanium white, or even titania) is cheap, nontoxic, readily available in bulk, and very widely used as a white pigment in paint, enamel, lacquer, plastic, and construction cement. TiO_2 powder is chemically inert, resists fading in sunlight, and is very opaque: This allows it to impart a pure and brilliant white color to the brown or gray chemicals that form the majority of household plastics. In nature, this compound is found in the minerals anatase, brookite, and rutile.

Paint made with titanium dioxide does well in severe temperatures, is somewhat self-cleaning, and stands up to marine environments. Pure titanium dioxide has a very high index of refraction and an optical dispersion higher than diamond. Star sapphires and rubies get their asterism from the titanium dioxide present in them. Titanates are compounds made with titanium dioxide. Barium titanate has piezoelectric properties, thus making it possible to use it as a transducer in the interconversion of sound and electricity. Esters of titanium are formed by the reaction of alcohols and titanium tetrachloride and are used to waterproof fabrics.

Titanium nitride is often used to coat cutting tools, such as drill bits. It also finds use as a gold-colored decorative finish, and as a barrier metal in semiconductor fabrication.

Titanium (IV) chloride (titanium tetrachloride, $TiCl_4$, sometimes called “Tickle”) is a colourless, weakly acidic liquid which is used as an intermediate in the manufacture of titanium(IV) oxide for paint. It is widely used in organic chemistry as a Lewis acid, for example in the Mukaiyama aldol condensation. Titanium also forms a lower chloride, titanium(III) chloride ($TiCl_3$), which is used as a reducing agent.

Titanocene dichloride is an important catalyst for carbon-carbon bond formation. Titanium isopropoxide is used for Sharpless epoxidation. Other compounds include; Titanium bromide (used in metallurgy, superalloys, and high-temperature electrical wiring and coatings) and titanium carbide (found in high-temperature cutting tools and coatings).

Applications

About 95 percent of titanium production is consumed in the form of titanium dioxide (TiO_2), an intensely white permanent pigment with good covering power in paints, paper, toothpaste, and plastics. Paints made with titanium dioxide are excellent reflectors of infrared radiation and are therefore used extensively by astronomers and in exterior paints. It is also used in cement, in gemstones, as an optical opacifier in paper (Smook 2002), and a strengthening agent in graphite composite fishing rods and golf clubs. Recently, it has been put to use in air purifiers (as a filter coating), or in film used to coat windows on buildings which when exposed to UV light (either solar or man-made) and moisture in the air produces reactive redox species like hydroxyl radicals that can purify the air or keep window surfaces clean.

Watch with titanium cover.

Because of its high tensile strength (even at high temperatures), light weight, extraordinary corrosion resistance, and ability to withstand extreme temperatures, titanium alloys are used in aircraft, armor plating, naval ships, spacecraft, and missiles. It is used in steel alloys to reduce grain size and as a deoxidizer, and in stainless steel to reduce carbon content. Titanium is often alloyed with aluminum (to refine grain size), vanadium, copper (to harden), iron, manganese, molybdenum, and with other metals.

Welded titanium pipe is used in the chemical industry for its corrosion resistance and is seeing growing use in petroleum drilling, especially offshore, for its strength, light weight, and corrosion resistance.

Titanium alloyed with vanadium is used in the outer skin of aircraft, fire walls, landing gear, and hydraulic tubing. An estimated 58 tons of the metal is used in the Boeing 777, 43 in the 747, 18 in the 737, 24 in the Airbus A340, 17 in the A330 and 12 in the A320, according to the 2004 annual report of the Titanium Metals Corporation. Generally, newer models use more and widebodies use the most. The A380 may use 77 tons, including about 10 or 11 tons in the engines.

Use of titanium in consumer products such as tennis rackets, golf clubs, bicycles, laboratory equipment, wristwatches, wedding bands, and laptop computers is becoming more common.

Other Uses

- Due to excellent resistance to sea water, it is used to make propeller shafts and rigging and in the heat exchangers of desalination plants and in heater-chillers for salt water aquariums, fishing line and leader, and lately diver knives as well.
- Owing to its strength and inertness to seawater, as well as its substantial ore deposits in Russia, it was the principal material used in the construction of many advanced Russian submarines, including deepest-diving military submarines to date, Alfa and Mike class, as well as Typhoon class.
- It is used to produce relatively soft artificial gemstones.
- Titanium tetrachloride ($TiCl_4$), a colorless liquid, is used to iridize glass and because it fumes strongly in moist air it is also used to make smoke screens and in skywriting.
- In addition to being a very important pigment, titanium dioxide is also used in sunscreens due to its resistance to UV Radiation.
- Because it is considered to be physiologically inert, the metal is used in joint replacement implants such as hip ball and sockets and to make medical equipment and in pipe/tank lining in food processing. Since titanium is non-ferromagnetic, patients with titanium implants can be safely examined with magnetic resonance imaging (convenient for long-term implants).
- Titanium is also used for the surgical instruments used in image-guided surgery.
- Titanium blades are sometimes used in movies. Titanium ripping against titanium will produce blue, spectacular lightning and forging such weapon (which is hard and very expensive) may still end up cheaper than equivalent digital post-processing, especially for heroic fantasy movies.

- Its inertness and ability to be attractively colored makes it a popular metal for use in body piercing.
- Titanium has the unusual ability to osseointegrate, enabling use in dental implants. This ability is also exploited by some orthopaedic implants. Orthopaedic applications also take advantage of titainium's lower modulus of elasticity to more closely match the modulus of the bone that such devices are intended to repair. As a result, skeletal loads are more evenly shared between bone and implant leading to a lower incidence of bone degradation from stress shielding and periprosthetic bone fractures which occur at the boundaries of orthopaedic implants which act as stress risers. However, titanium alloys' stiffness is still more than twice that of bone, eventually leading to joint degradation.
- Titanium alloys are also used in spectacle frames. This results in a rather expensive, but highly durable and long lasting frame which is light in weight and causes no skin allergies. Both traditional alloys and shape memory alloys find use in this application.
- Many backpackers use titanium equipment, including cookware, eating utensils, lanterns and tent stakes. Though slightly more expensive than traditional steel or aluminum alternatives, these titanium products can be significantly lighter without compromising strength.
- Titanium cookware is available, and often more expensive than traditional cookware based on iron, copper, or aluminum. However, the thermal properties of titanium cookware can cause uneven heating without a well-distributed heat source, which may make it unsuitable for some culinary applications.
- Titanium has a high strength-to-weight ratio and shock-absorbing properties similar to steel alloys when used to manufacture high-quality bicycle frames. Titanium has a higher cost but has the advantage of corrosion resistance.
- Titanium is increasingly used in lacrosse stick shafts.
- Titanium is increasingly being used in cricket helmet grills.
- Titanium may be anodised to produce various colours.
- Titanium is also present in fireworks.
- Titanium comes in the form of foil, sheet, wire, granules, sponge, nanosized activated powder, powder, mesh, and rod.

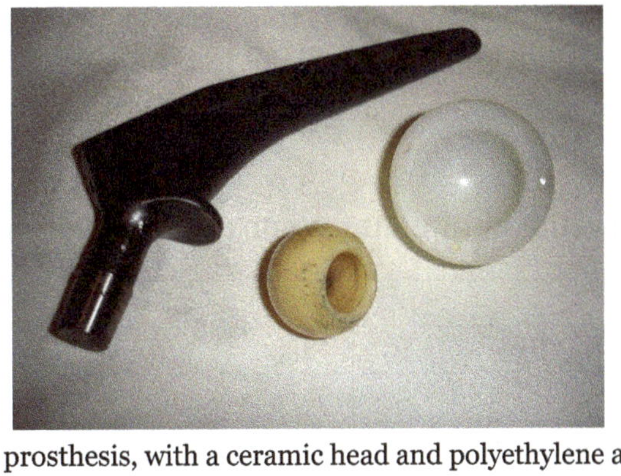

A titanium hip prosthesis, with a ceramic head and polyethylene acetabular cup.

Titanium has occasionally been used in construction: The 150-foot (45 m) memorial to Yuri Gagarin, the first man to travel in space, in Moscow, is made of titanium for the metal's attractive colour and association with rocketry. The Guggenheim Museum Bilbao and the Cerritos Millennium Library were the first buildings in Europe and North America, respectively, to be sheathed in titanium panels.

Precautions

As a powder or in the form of metal shavings, titanium metal poses a significant fire hazard and, when heated in air, an explosion hazard. Water and carbon dioxide-based methods to extinguish fires are ineffective on burning titanium; Class D dry powder firefighting agents must be used instead.

Salts of titanium are often considered to be relatively harmless but its chlorine compounds, such as $TiCl_2$, $TiCl_3$ and $TiCl_4$, have unusual hazards. The dichloride takes the form of pyrophoric black crystals, and the tetrachloride is a volatile fuming liquid. All of titanium's chlorides are corrosive. Titanium also has a tendency to bio-accumulate in tissues that contain silica but it does not play any known biological role in humans.

Titanium Alloys

Though a large number of titanium alloys have been developed, they can be grouped into three major categories such as α-alloys, β-alloys and α+β alloys. Their properties are dependent on microstructure which, in turn, depends on the chemical composition and thermomechanical processing.

Due to their high specific strength and exceptional corrosion resistance, titanium alloys have been widely used in engineering, namely in the aerospatiale, automotive and biomedical fields. The alloy Ti-6Al-4V, due to its biocompatibility coupled with good combination of mechanical and corrosive properties, is the most popular titanium- based material in the manufacture of biomedical components, including dental implants, prosthetic femoral components, surgical instruments, etc. Nowadays, the industry of titanium alloys seems to be mature, but new technology and applications for these alloys continue to grow.

Regardless of the usefulness of titanium alloys, the number of available articles addressing the subject has been limited. The aim of this paper is to perform a review on titanium alloys. The core subjects are properties and applications, with special focus on Ti-6Al-4V. Additional descriptions on elemental titanium, alloying elements, chemical composition, and classification are also provided.

Table: Selected physical properties of titanium, compared to some competitor metals.

	Ti	Al	Fe	Ni
Density [g/cm³]	4.5	2.7	7.9	8.9
Melting point [°C]	1670	660	1538	1455
Thermal conductivity [W/mK]	15-22	221-247	68-80	72-92

Elastic modulus [GPa]	115	72	215	200
Reactivity with oxygen	high +	high	low	low
Corrosion resistance	high +	high	low	medium
Metal price	high +	medium	low	high

Elemental Titanium and Alloying Elements

Elemental Titanium

Elemental or pure titanium has low thermal conductivity, relatively low density and elastic modulus, moderate strength, good corrosion resistance in various environments and high reactivity with a variety of elements. Table shows selected properties of this element compared with those of some competitor metals.

At low temperatures, pure titanium has hexagonal close packed structure (hcp), called titanium. But at high temperatures the stable structure is body-centered cubic (bcc), which is referred to as titanium. The atomic unit cells of the referred structure are illustrated in figure. The –β transus temperature for pure titanium is around 882±2 °C and is dependent on the amount of incorporated impurities. The existence of the two different crystal structures is the basis for large variety of titanium alloys properties.

Alloying Elements

The alloying elements for titanium alloys are classified as neutral, α-stabilizers or β-stabilizers, according to their influence on stabilizing or phases. This table also evidences the position of alloying element in the lattice crystal, which may be interstitial or substitutional.

The α-stabilizers rise the β-transus temperature, while -stabilizers lower this temperature. Neutral elements have slight influence on β-transus temperature. On the other hand, β-stabilizers are divided into β-isomorphous and β-eutectoid elements. The first promote the stability of phase along all composition of the alloy, while the last cause eutectoid transformations of phase.

Aluminum is the most important α-stabilizer, while molybdenum is among the main β-stabilizers. The α-stabilizers and their relative capacity to stabilize phase are expressed as aluminum equivalence, having the molybdenum similar meaning for phase.

Properties of Titanium Alloys

The mechanical properties of titanium alloys are influenced by individual properties of αand β phases, their arrangement and volume fraction. For example, the α phase has lower density than the β phase due to the fact that the predominant element Al in a phase has lower density than the predominant elements Mo or V in β phase. Consequently, αalloys generally have lower density than β alloys, but the values are about 4.5 g/cm^3.

Besides, the fine microstructures increase strength and ductility, retard crack nucleation and are a pre-requisite for super plastic deformation, while coarse microstructures are more resistant to creep and fatigue crack growth; equiaxed microstructures often have high ductility and fatigue strength and are ideal for super plastic deformation, whereas lamellar structures have high fracture toughness and show superior resistance to creep and fatigue crack growth; bimodal microstructures combine the advantages of lamellar and equiaxed structures, which results in a well-balanced property profiles.Table provides selected mechanical properties of some of the most used titanium alloys. The chemical composition and β-transus temperatures are also indicated.

Properties of Ti-6Al-4V Alloy

The classic Ti-6Al-4V alloy is employed generally in two metallurgical conditions: annealed and solution treated and aged. Table provides the magnitude of selected properties of this alloy.

Figure provides information on the Influence of temperature on properties of Ti-6Al-4V alloy. The thermal conductivity increases to more than threefold when the temperature changes from about 10 °C to about 1200 °C while the hardness increases rapidly as the temperature is decreased. Hong et al. underlined that when temperature changes from room to liquid nitrogen temperature, tensile strength of this alloy changes from 1000 to 1700 MPa. There is an overall decrease of Young modulus with increasing temperature from above 200 °C but this decrease is more pronounced for temperatures above 500 °C. The ultimate tensile strength (UTS) decreased from about 1000 MPa at 25 °C to about 625 MPa at 450 °C which results in a global decrease equal to about 45%.

The elongation increases with increasing volume fraction of phase, thickness of phase and the Feret ratio. The correlation of microstructure and mechanical properties was developed using the artificial neural network technique, being the mean absolute percentage error only 1.22%, which imply that the trained neural network is able to predict the mechanical properties of the Ti-6Al-4V alloy accurately.

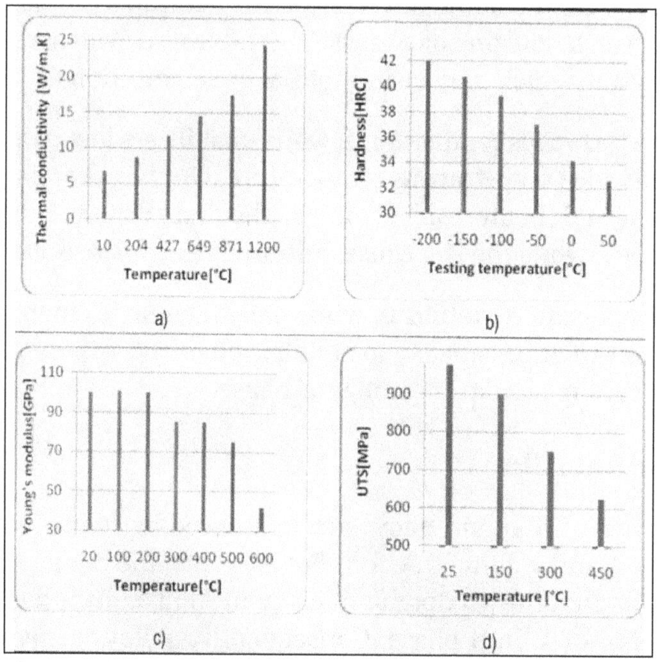

Influence of temperature on selected properties of Ti-6Al-4V alloy.

Figure provides flow stress magnitude versus strain, obtained at room temperature (300K), for strain rate equal to 1500 S-1. The flow stress increases with increasing strain. This behavior is usually caused by hardening phenomenon. According to figure, the flow stress increases with strain rate. Besides, high hydrogen content in Ti-6Al-4V alloy resulted in increase of flow stress.

Figure presents flow stress magnitude of Ti-6Al- 4V alloy versus temperature, for different combinations of hydrogen content and strain rate. The increase of temperature leads the decrease of flow stress magnitude which, however, generally increased with increasing hydrogen content and/ or strain rate, except for temperature of 500 °C.

Constitutive analysis of Ti-6Al-4V alloy was developed by performing hot compression tests in a wide range of temperature (800-1050 °C) and strain rates (0.0005-1 s^{-1}). Strain showed significant influence on material constants. The developed constitutive equation can precisely predict the flow stress under most deformation conditions in α+β phase region. Almost similar prediction accuracy was also obtained in single β phase region. However, some significant deviations are observed between the experimental and the computed flow stress data for some deformation conditions in α+β phase region. Almost similar prediction accuracy was also obtained in single phase region. However, some significant deviations are observed between the experimental and the computed flow stress data for some deformation conditions in β phase region.

Mechanical tests performed at room temperature revealed that the hydrogen content (0.2-1.2 wt.%) has dissimilar effect on the tensile and compressive properties of Ti-6Al-4V alloy. As hydrogen content increases, the tensile properties decrease, but the compressive properties first increase and then decrease. The decrease of tensile properties with increasing hydrogen content resulted from increasing amounts of hydrides and solid solution of hydrogen at the grain and phase interfaces, which causes the decrease of the binding force at the referred interfaces. The increase of hydrogen content also causes an increase in the amount of plastic , which result in increase of the ultimate compression strength. However, high hydrogen content in the specimen results in high amounts of hydrogen content in the form of solid solution and hydrides, which are continuously distributed along the grain boundaries and leads to the degradation of compressive properties. The optimum hydrogen content for cold forming of Ti-6Al-4V alloy for the highest ultimate compression strength was found to be in the range of 0.6-0.8 wt.%.

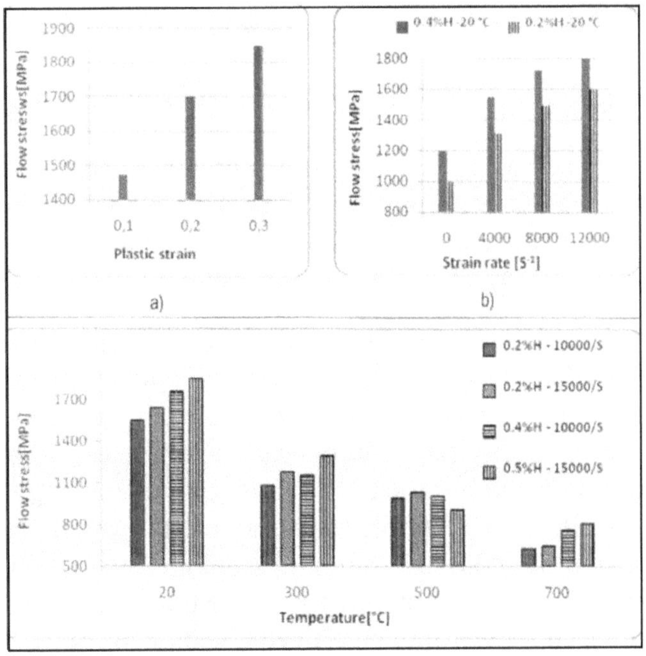

(a) Influence of plastic strain, (b) strain rate and hydrogen content, (c) temperature and hydrogen content on the evolution of flow stress of Ti-6Al-4V alloy.

The effect of hydrogen on the cold deformation behaviors of the Ti-6Al-4V alloys (α+β type) was investigated. Microstructures exhibiting large α" martensite and metastable β phase show a better plasticity than that of the as received alloy. The deformation twins appear inside of α" martensite in specimens containing lower than 0.45 wt.% H, which promotes the deformation. When the hydrogen concentration is further increased up to 0.9 wt.% the metastable β phases transform to α" martensite during the deformation, which, together with the deformation twins, increase the true strain from 30% to 60%.

The oxygen content also influences the mechanical properties of titanium alloys. deoxidation of Ti-6Al-4V alloy results in microstructure refinement and decrease of microhardness.

Figure provides nanohardness values of nitrogen ion implanted versus unimplanted Ti-6Al-4V alloy. The nanohardness values were obtained from experimental test using the applied loads of 2 mN and based on load-displacement curves. Nitrogen ion implantation caused an increase of about 20% in nanohardness. This growth was attributed by Luo to higher stiffness of hard layer of TiN formed on implanted Ti-6Al-4V substrates.

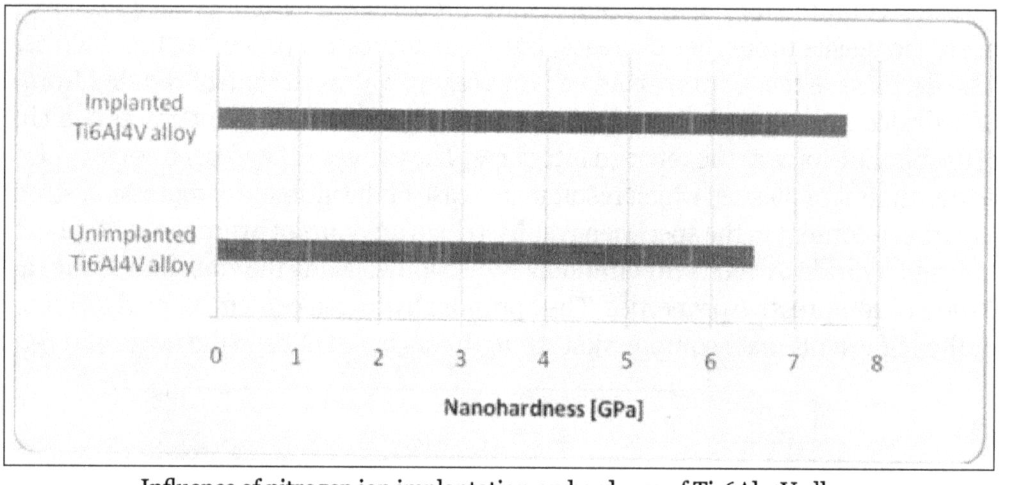

Influence of nitrogen ion implantation on hardness of Ti-6Al-4V alloy.

Porous materials are often used for biomedical parts and device construction. They investigated the formation of porous morphology in nanocrystalline of Ti-6Al-4V alloy. The results show that, due to the porous morphology, the nanocrystalline alloy has a nanohardness and Young modulus in the range of993-1275 HV and 137-162 GPa, respectively.

Poondla measured the microhardness (Vickers scale) and macrohardness (Rockwell C scale) of commercially pure titanium (Grade 2) and Ti-6Al-4V alloy, both in the annealed condition. Marginal variation of microhardness through a cross- section of the samples was found and the authors attributed this behavior to the change in microstructure, namely the αand β phases, being the average values about 200 kg/mm² (1962 MPa) and 330 kg/mm² (3237.3 MPa), respectively for Grade 2 and Ti-6Al-4V alloy. The macrohardness across the length of machined sample gave an average value 260 kg/mm² (2550.6 MPa) for Grade 2 and 290 kg/mm² (2844.9 MPa) for Ti-6Al-4V. The microhardness values were marginally lower than the macrohardness values across the length of Grade 2 sample, but regarding the Ti-6Al-4V alloy the opposite happened. The lower value of macrohardness for Ti-6Al-4V alloy when compared to its microhardness was attributed by the authors to the physical presence of a population of processing-related artifact.

Study of Zherebtsov showed that the application of hydrostatic extrusion and multiaxial forging followed by cold rolling results in Ti-6Al-4V alloy with very high tensile strength of about 1500 MPa, while the multiaxial forging, and forging combined with heat treatment results in Ti-6Al-4V alloys with moderate strengths of about 1300 MPa and 1050 MPa, respectively.

Cvijovic et al. studied the wear and corrosion behavior in Ringer's solution. The materials investigated were the cold-rolled Ti-13Nb-13Zr alloy with martensitic microstructure, and Ti-06Al-4V ELI alloy in martensitic and two-phase (α+β) microstructural conditions. The wear experiments were carried out at room temperature with a normal load of 40 N and sliding speeds 0.26, 0.5, and 1.0 m/s. The corrosion behavior was analyzed at 37 °C using open circuit potential-time measurements and potentiodynamic polarization. It was concluded that Ti-6Ai-4V ELI alloy has a substantially higher wear resistance than Ti-13Nb-13Zr alloys in both microstructural conditions. Both alloys exhibited spontaneous passivity in Ringer's solution and the corrosion potential values are similar for all three materials. However, the corrosion resistance of Ti-13Nb-13Zr and martensitic Ti-6Al-4V ELI alloy presented the best combination of both corrosion and wear resistance.

The alloy Ti-6Al-4V has the ability to substitute heavier steel in some friction and wear-critical diesel engine components like connecting rods, intake valves, movable turbocharger vanes, and pistons. This alloy also exhibits excellent corrosion resistance, good fatigue strength, and acceptable fracture toughness, but it has poor sliding characteristics. In addition, titanium alloys have a propensity to fail by galling, and often exhibit high and unstable friction coefficients. Selected surface engineering techniques were compared to determine which best enhance the tribological performance of Ti-6Al-4V alloy. Candidate treatements included diffusion treatments (oxygen diffusion, nitriding, and carburizing), hard coatings (TiN and CrN), soft coating (Cu-Ni-In) titanium-matrix TiB2 in situformed composite and shot peening. One of the interesting conclusions is that the CrN coating had the least amount of wear.

Study of micromechanical behavior of Ti-6Al-4V alloy through micromechanical behavior of Ti-6Al-4V alloy through microindentation experiments revealed that the loading speed has little influence on microhardness and Young's modulus, but showed strong indentation size effects, namely an increase of indentation hardness with the decrease of indentation load or depth.

The surfaces of Ti-6Al-4V alloy component modified via CoBlast using either Al_2O_3-SiC or Al_2O_3-B_4C powder mixtures demonstrated increased hardness compared with the unmodified alloy, but wear testing performed using a pin-on-disc tribometer revealed that this treatment did not cause significant improvement in wear resistance.

The application of titanium alloys as implant in the bone is usually very good, but has certain restrictions because the metal ions are released from the implant and may cause local irritation of the surrounding tissues. Cell and tissue are affected not only by the chemical properties of the implant surface, but also by the implant surface topography. To overcome the problem of ion release and to improve the biological, chemical, and mechanical properties, many surface treatment techniques are used. calcium and phosphorus implantation are useful for improvement of biocompatibility of titanium. They also describe that the silver ion implantation is used for anti- bacterial applications, while nitrogen ion implantation combats wear of the titanium surface.

Titanium and titanium alloys are characterized by excellent mechanical properties, high toughness, high strength and relatively low weight. Moreover titanium shows great corrosion resistance in many environments, high biocompatibility, and also some titanium alloys show good heat resistance. Thus, the combination of these properties and characteristics of titanium alloys have led to successful applications which demand high levels of reliable performance mainly in aerospace, military, automotive, chemical and power plant, petroleum and also in surgery and medicine, as well as in many other major industries.

At the same time titanium alloys have disadvantages. The most important are a high cost, poor tribological properties such as poor abrasive wear resistance, relatively low hardness and high coefficient of friction. The low surface hardness and low wear resistance limit the application of titanium alloys under conditions of friction and contact loads. Therefore various methods for improving the surface properties and tribological characteristics of titanium alloys are applied. Tribological properties of titanium alloys may be improved or modified by coatings created on the surface, changing microstructure of surface layers or thermochemical treatments such as nitriding, carburizing, oxidizing or nitrooxidizing, etc.

The most commonly used methods of surface modification are thermal spraying of coatings, cladding, physical vapour deposition (PVD), chemical vapour deposition (CVD), saltbath nitriding, ion implantation or plasma nitriding. These methods have been developed and applied for many years.

In the case of thermal spraying and other methods of adhesive layers producing, the crucial parameter, decisive for quality and successful application, is adhesion of the bonding between metallic or ceramic coating and the titanium substrate. However during plasma spraying an oxide layer may be formed on the titanium metallic surface, which is very harmful and may be the reason of adhesion loss between the coating and metallic substrate. In turn, nitriding of surface layers of titanium alloys is one of most widely used thermochemical treatments in the industry but all the processes of surface nitriding mentioned above have serious disadvantages. For example plasma nitriding requires several hours of processing time. Moreover some of the nitriding processes are performed at high temperatures, therefore the bulk of the nitrided sample or a part is exposed to high temperature, so the microstructure and mechanical properties of the bulk material can be also changed and additionally non-uniformity of surface microstructure can occur. Additionally, a serious disadvantage of the TiN coatings is poor ductility and significant reduction in fatigue wear resistance. In a case of hard coatings composed of titanium nitrides TiN deposited by PVD, CVD and plasma directly on the titanium or titanium alloy substrate, a failure at the interface between the coating and the substrate is very likely. One of the most promising and efficient methods of surface modification of titanium alloys is laser gas nitriding (LGN).

Since the first trials of laser gas nitriding of titanium by Katayama in the eighties of the last century, there have been many studies on the process of laser nitriding of titanium alloys, but the studies were performed under laboratory conditions. Moreover the technology is not applied in the industry yet.

Introduction of the new generation of high power direct diode lasers (HPDDL) to industry allows for further development of the laser surface modification of titanium alloys.

Results of previous investigations on the nitriding of titanium by HPDDL lasers show an advantage of the diode lasers over other types of lasers used for laser gas nitriding . Therefor the detailed study of tribological characteristic of composite surface layers of titanium alloy Ti6Al4V produced by direct diode laser gas nitriding were undertaken.

Titanium Alloy Classification

Titanium alloys are typically classified as pure titanium, alpha, beta and alpha-beta alloys. There are also so-called near-alpha-phase and near-beta-phase (i.e. metastable beta) alloys. Under equilibrium conditions, pure titanium and alpha (a) phase have hexagonal-closed-packed (HCP) structures up to 882°C (1620°F). Above this temperature, they transform to beta (b) phase having a body-centered-cubic (BCC) structure up to the alloy's melting point. Near-alpha alloys typically have a small amount of beta-phase (1-2%) stabilizing elements. In near-beta alloys, significant additions of beta stabilizer suppress the M_s temperature to below ambient, and the beta phase is retained at room temperature by rapid cooling or quenching from the alpha-beta phase. The inherent properties of all these structures are quite different.

Alloying elements generally stabilize one or the other of these phases. The alpha phase is stabilized by aluminum, gallium, germanium, boron, carbon, oxygen and nitrogen, while the beta phase is stabilized by molybdenum, vanadium, tantalum, niobium, iron and hydrogen.

The boundary between the alpha and beta phase and the two-phase alpha-beta region is called the alpha transus and beta transus respectively. The beta-stabilized system has M_s and M_f temperatures associated with it.

The titanium grades most commonly used have compositional specifications determined by ASTM E120 (Standard Test Methods for Chemical Analysis of Titanium and Titanium Alloys) for the commercially important alloys. Military specifications are found under MIL-T-9046 and MIL-T-9047, and aerospace material specifications for bar, sheet, tubing and wire are under AMS specification numbers 4900-4980. In addition, large aircraft companies have their own set of alloy specifications. Because titanium alloys are used in a variety of applications, several different material and quality standards are specified. Among these are ASTM, ASME, AMS, the U.S. military and a number of proprietary sources.

Titanium alloys may be divided into two principal categories: Corrosion resistant and structural. The corrosion-resistant alloys are generally based on a single-phase (alpha) with dilute additions of solid-solution strengthening and alpha-stabilizing elements such as oxygen (interstitial), palladium, ruthenium and aluminum (substitutional). These alloys are used in the chemical, energy, paper and food-processing industries to produce highly corrosion-resistant tubing, heat exchangers, valve housings and containers. The single-phase alpha alloys provide excellent corrosion resistance, good weldability, and easy processing and fabrication but a relatively low strength.

The beta phase is stabilized by additions such as molybdenum, vanadium, niobium, iron (substitutional) and hydrogen (interstitial). A dispersion of alpha in the beta matrix along with solid-solution strengthening of both the alpha and beta phases lead to higher-strength alloys referred to as structural alloys. The structural alloys can be divided into four categories: the near-alpha alloys, the alpha-beta alloys, the beta alloys and the titanium aluminide (ordered) intermetallics (based on Ti_xAl where x = 1 or 3). With titanium alloys used in structural applications, optimization of mechanical properties is very important. Therefore, processing and microstructure control are critical.

Heat Treatment of Titanium Alloys

Basics of Heat Treatment

Figure illustrates the effect of alloying elements on α, and or β in alloys. It also shows that the strength of annealed alloys increases linearly and gradually with solute content, or as amount of β-phase increases. Quenched alloys transform β to martensite.

The resulting effect in increasing strength is not much except at low concentration of solute. The maximum increase in strength occurs by age-hardening by decomposing retained β-phase. Two phase alloys are stronger than single alpha phase as beta phase is stronger than alpha phase.

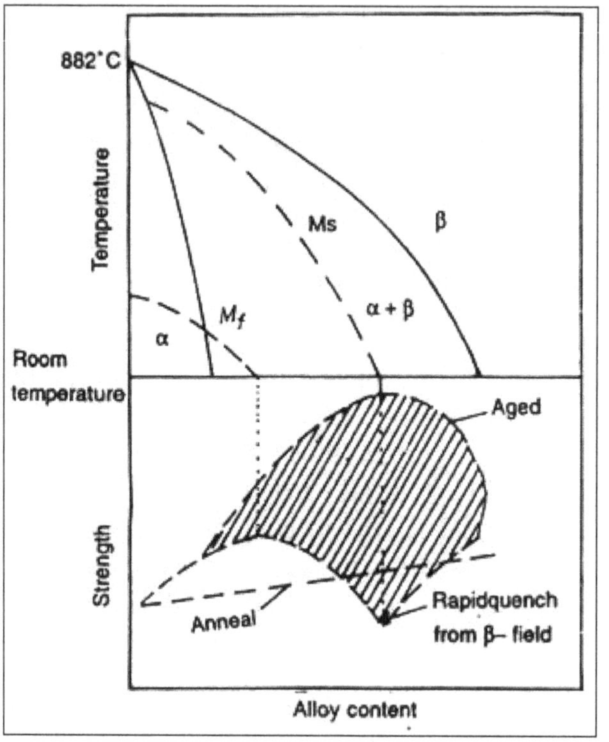

Schematic heat treatment of titanium allouys.

The alloys having α and nearly-α (β max 2%) are given stress-relieving and annealing heat treatment. The two-phase alloys containing α + β show maximum response to hardening and strengthening.

Heat Treatment Processes for Titanium Alloys

- Stress relieving to relieve residual stresses induced during fabrication.
- Annealing to have maximum ductility, machinability. The stable micro-structure results in dimensional stability also.
- Solution heat treatment and precipitation hardening to improve the mechanical strength and hardness.

Stress Relieving

It is the process of heating to a high temperature, soaking at this temperature to relieve stresses without causing an undesirable amount of precipitation, or strain ageing in α+ β and β-type alloys, or undesirable recrystallisation in single phase alloys which depend on cold work for strength and hardness.

Stress relieving does not cause adverse effect on strength, or ductility. A separate stress relieving treatment may be omitted when the sequence in production employs annealing or hardening.

Table: Stress relieving temperature and time range for titanium alloys.

Alloy	Temperature Range	Time hrs.
1. Ti (Commercially pure)	480-590 °C	1/4-4
2. α-Titanium alloys:		
Ti-5 Al-5 Sn-% Zr	595-705	1/4-1
Ti-8 Al-1 Mo-1 V	595-705	1/2-5
Ti-8 Al-2 Cbl Ta	595-705	1/4-1
3. α+β alloys:		
Ti-2 Fe-2 Cr-2 Mo	480-540	1/2-1
Ti-4 Al-4 V	480-650	1-50
Ti-6 Al-6 V	480-650	1-50
Ti-7 Al-4 Mo	480-700	1-8
4. B-alloys:		
Ti-13 V-11 Cr-3 Al	700-730	1/12-1/4

Annealing

Quite a few components of titanium alloys are used in service in the annealed state.

The aims of annealing are:

- To induce maximum ductility at room temperature.
- To improve machinability.
- To induce adequate toughness.
- To induce dimensional and microstructural stability at high temperatures i.e., to induce maximum creep resistance.

Thermal stability is obtained by producing stable beta, which is capable of resisting further transformation, when' exposed to elevated temperatures in service. Annealing should be able to obtain this beta-phase from, if formed, brittle intermediate omega phase. The annealing process consists of heating to 650° to 790 °C (depending on the composition), soaking and then slow cooling (3 °C/min), to 540 °C, and cooling in air.

'Duplex' annealing helps to obtain maximum creep resistance in alloys such as Ti-8 Al- 1 Mo – 1 V. It consists of heating the alloy in range of 790°-900 °C for a time of 5 minutes for sheets and to 1 hour for thicker bars, and then cooled in air. The alloy is then given 'stabilising anneal' by heating it to 480° to 700 °C. Table gives annealing temperatures for some titanium alloys:

Table: Annealing treatment of some titanium alloys.

Alloy	Heating Temperature, °C	Tune (hr.)	Post Soaking cooling
1. Pure Ti (Commercially pure)	650-760	1/10-8	Air
2. α-titanium alloys:			
Ti-5 Al-2.5 Sn	720-845	1/6-8	Air
Ti-8 AI-2 Cb-1 Ta	800-900	1-8	Air
Ti-8 Al-1 Mo-1 V	800-900	1 – 8	Air
3. α+β Titanium Alloys:			
Ti-4 Al-4 V	650-760	1 – 8	Air
Ti-2 Fe-2 Cr-2 Mo	650-700	1/2 -4	Air
Ti-6 Al-6 V	700-845	I – 8	Air/Fce
Ti-7 Al-4 Mo	700-790	I – 8	
4. β-Titanium Alloys			
Ti-13 V-11 Cr-3 Al	700-800	1/6-1	Air. Water
Ti-15 V-3 Al-3 Cr-3 Sn	790-810	1/12-1/4	Air

Precipitation Hardening

It is possible to strengthen several titanium alloys with adequate ductility by giving them precipitation hardening.

Solution Treatment

All the three types of titanium alloys are given precipitation hardening. The solution treatment is given in (α + β) region, which for commercial α + β alloys is 760 °C to 1000 °C. Changing the solution temperature alters the amount of β-phase. If solutionising is done in 100%, β-phase region, the resulting ductility is impaired. During ageing, it is the β-phase which decomposes to yield high strength levels. Thus, depending on the temperature of solution treatment, the amount of β-phase is varied (increases with temperature), which causes resulting age hardening with required ductility.

Quenching

Normally β-alloys are air-quenched from solution temperature, (α + β) alloys are water-quenched, or 5% brine-quenched. (Table gives the solution and ageing temperatures of some alloys). Actually, cooling rate from solution temperature plays an important role in strengthening the alloys. A slower cooling allows the diffusion to occur during cooling itself the changed β phase may not provide effective strengthening during ageing.

However, rapid rates of quenching is less critical in many alloys, but an alloy Ti-6 Al-4 V should be cooled from 930 °C to 540 °C in less than 2 seconds to develop maximum strength after ageing. Section size directly affects the effectiveness of quenching in such cases. Alloy Ti-4 Al-3 Mo-1 V may be cooled in oil, and still provides good strength after ageing.

Ageing

Ageing decomposes the supersaturated beta-phase retained on quenching. Proper ageing produces high strength with adequate ductility. In α + β alloys, various combination of strength, ductility and fabricability can be obtained by modifying the solution and ageing treatments.

Table: Solution and ageing temperatures of some titanium alloys.

Alloy	Solution Treatment		Ageing Treatment	
	Temperature °C	Time hr	Temperature °C	Time hr
1. α-Titanium Alloy:				
Ti-8Al-1Mo-1 V	982-1000	1/12-1/4	560 - 620	7 - 9
2. α+β Titanium Alloys				
Ti-6A1-4V	900-940	1/12-1/4	480-540	3 - 5
Ti-7AI-4Mo	930-980	1/2-2	540-625	4 - 8
Ti-2Fe-2Cr-2Mo	760-800	1/2-1	480-510	2 - 8
3. β-Titanium Alloys				
Ti-13 V-11 Cr-3 Al Sheet	800	1/12-1/3	425-540	12 - 100
Ti-13V-11 Cr-3 Al forging	None	None	425-480	1/2-4

In (α + β) alloys, quenching results invariably in beta transforming to a metastable but, brittle phase omega severe quenching and rapid heating to high ageing temperature can avoid it. Ageing above 430 °C, can change omega (to β). Table gives ageing treatment for some titanium alloys.

The only 100% beta alloy Ti- 13V-11 Cr-3 Al which forged and rolled in beta range, and air cooled after hot working yields solution treated condition. This alloy when aged at 500 °C for 8 to 100 hrs. yields a UTS of 1100-1380 MN/m^2.

Beryllium

Beryllium (chemical symbol Be, atomic number 4) ranks among the lightest of all known metals. Steel-gray in color, it is strong but brittle. It is classified as an alkaline earth metal, along with calcium and magnesium. Precious forms of its mineral beryl are aquamarine and emerald. This metal, its alloys, and compounds are toxic and need to be handled with care. One needs to especially avoid inhaling their dust and vapors.

Alloys of beryllium with copper or nickel are strong and wear-resistant, and are widely used in the manufacture of such items as computer parts, gyroscopes, electrodes, springs, and nonsparking tools. In addition, its alloys are useful as lightweight structural materials for high-speed aircraft, missiles, spacecraft, and satellites. The nuclear power industry uses beryllium in nuclear reactors as a neutron reflector and moderator. In X-ray lithography, beryllium is used for the reproduction of microscopic integrated circuits. Beryllium oxide is useful for applications that require an excellent heat conductor and electrical insulator, with a very high melting point and high strength and hardness.

Occurrence and Production

Beryllium is a significant constituent of about 100 minerals. The most important of these are beryl ($Al_2Be_3Si_6O_{18}$), bertrandite ($Be_4Si_2O_7(OH)_2$), chrysoberyl (Al_2BeO_4), and phenakite (Be_2SiO_4). Precious forms of beryl are aquamarine and emerald.

Beryllium ore.

The main commercial sources of beryllium and its compounds are beryl and bertrandite. Beryllium metal did not become readily available until 1957. Currently, the chief method of production is by reducing beryllium fluoride with magnesium metal. The chemical reaction can be written in terms of the following equation.

$$BeF_2 + Mg \rightarrow MgF_2 + Be$$

Physical properties	
Phase	solid
Density (near r.t.)	1.85 g/cm³
Liquid density at m.p.	1.690 g/cm³
Melting point	1560 K (1287 °C, 2349 °F)
Boiling point	2742 K (2469 °C, 4476 °F)
Heat of fusion	7.895 kJ/mol
Heat of vaporization	297 kJ/mol
Heat capacity	(25 °C) 16.443 J/(mol·K)

Vapor pressure						
P/Pa	1	10	100	1 k	10 k	100 k
at T/K	1462	1608	1791	2023	2327	2742

Notable Characteristics

Beryllium is a member of the family of alkaline earth metals and lies at the top of group two (former group 2A) of the periodic table, just above magnesium. In addition, it is situated in period two, between lithium and boron. Among the light metals, beryllium has one of the highest melting points. Its modulus of elasticity (a measure of its elasticity) is approximately one-third more than that of steel. It is nonmagnetic and an excellent conductor of heat. The speed of sound in beryllium is about 12,500 meters per second—greater than that through any other element. Highly permeable to X-rays, it liberates neutrons when struck by alpha particles.

Beryllium resists attack by concentrated nitric acid. In addition, at standard temperature and pressure (0 °C, 100 kPa), it resists oxidation when exposed to air. It appears, however, that its ability to scratch glass may be due to the formation of a thin layer of its oxide.

Atomic properties	
Crystal structure	hexagonal
Oxidation states	2 (amphoteric oxide)
Electronegativity	1.57 (Pauling scale)
Ionization energies (more)	1st: 899.5 kJ/mol
	2nd: 1757.1 kJ/mol
	3rd: 14848.7 kJ/mol
Atomic radius	105 pm
Atomic radius (calc.)	112 pm
Covalent radius	90 pm

Isotopes

Beryllium has ten isotopes, of which only ^{9}Be is stable. The isotope ^{10}Be is produced in the atmosphere by the impact of cosmic rays on oxygen and nitrogen. This beryllium (being soluble in water at pH below 5.5) readily dissolves in rainwater (which generally has a pH less than five) and is transported to the Earth's surface. As this water quickly becomes more alkaline, ^{10}Be drops out of solution and accumulates at the soil surface. Given its half-life of 1.51 million years, ^{10}Be has a long residence time before it decays to ^{10}B (an isotope of boron). The isotope ^{10}Be and its daughter products have therefore been used to examine soil erosion, soil formation from regolith, and the development of lateritic soils. It has also been used to check variations in solar activity and the age of ice cores.

A square beryllium foil mounted in a steel case is ready to be used as a window between a vacuum chamber and an X-ray microscope. Beryllium filters out visible light but is highly transparent to X-rays.

The instability of the isotopes ^7Be and ^8Be has profound consequences for cosmological theory. It means that elements heavier than beryllium could not have been produced by nuclear fusion at the time of the Big Bang. Moreover, the nuclear energy levels of ^8Be are such that carbon can be produced within stars, thus making it possible for the eventual development of living organisms.

The shortest-lived isotope of beryllium is ^{13}Be, which has a half-life of 2.7×10^{-21} seconds and decays through neutron emission. The isotope ^6Be also is also very short-lived, with a half-life of 5.0×10^{-21} seconds.

Health Effects

Beryllium, its alloys, and its salts are toxic, and one should especially avoid inhaling their dust and fumes. Exposure to these materials can lead to two types of conditions: acute beryllium disease (acute berylliosis) and chronic beryllium disease (chronic berylliosis). In addition, long-term exposure can increase the risk of developing lung cancer.

An acute condition is that which develops after a short, heavy exposure and lasts for no more than a year. Depending on the type of exposure, acute beryllium disease may involve inflammation of the skin (contact dermatitis), nose and throat (nasopharyngitis), windpipe (tracheobronchitis), or lungs (pneumonitis).

Chronic beryllium disease develops after prolonged (months or years) of exposure to beryllium. The time lapse between first exposure and symptoms of the disease may be as much as ten to 15 years. The chronic disease affects mainly the lungs but its effects may also be seen in other organs. The condition produces lesions and granulomas (tissue masses) mainly in the lungs, but it may also affect other organs, such as the skin, liver, kidney, heart, bone, lymphatic system, and nervous system. Symptoms of the chronic disease include coughing, chest pain, and breathing difficulties.

Acute chemical pneumonitis was first reported in Europe in 1933 and in the United States in 1943. Cases of chronic berylliosis were first described in 1946 among workers in plants manufacturing fluorescent lamps in Massachusetts, and the use of beryllium compounds in fluorescent lighting tubes was discontinued in 1949. The potential for exposure to beryllium exists in the nuclear and aerospace industries, the refining of beryllium metal, melting of beryllium-containing alloys, the manufacture of electronic devices, and the handling of other beryllium-containing materials.

Early researchers tasted beryllium and its various compounds for sweetness, to verify its presence. Modern diagnostic equipment no longer necessitates this highly risky procedure, and no attempt should be made to ingest this substance. Beryllium and its compounds should be handled with great care, and special precautions must be taken when carrying out any activity that may involve the release of beryllium dust. No attempt should be made to work with beryllium before familiarization with correct handling procedures.

A successful test for beryllium on different surface areas has been recently developed. This technique, in which beryllium is bound to a fluorescent chemical (sulfonated hydroxybenzoquinoline), enables the detection of beryllium at concentrations up to ten times below the recommended limit for the workplace.

Industrial Release Limits

Typical levels of beryllium that industries may release into the air are on the order of 0.01 micrograms per cubic meter (μg/m³), averaged over a 30-day period, or two μg/m³ of workroom air for an eight-hour work shift. The current U.S. Occupational Safety and Health Administration (OSHA) permissible exposure limit for beryllium is two μg/m³, but this figure has been challenged as inadequate for protecting workers from developing beryllium sensitization and CBD. The American Conference of Governmental Industrial Hygienists (ACGIH), an independent organization of experts in the field of occupational health, has proposed a threshold limit value (TLV) of 0.05 μg/m³, in a 2006 Notice of Intended Change (NIC). As it can be difficult to control industrial exposures to beryllium, it is advisable to use any methods possible to reduce airborne and surface contamination by beryllium, to minimize the use of beryllium and beryllium-containing alloys whenever possible, and to educate people about the potential hazards if they are likely to encounter beryllium dust or fumes.

Uses of Beryllium

Because of its unmatched combination of qualities, beryllium has become an increasingly important material for a wide range of commercial and governmental applications.

Communications

Beryllium is used in the manufacture of telecommunications infrastructure equipment, computers and cellular phones, thereby helping people around the world to keep in touch.

People worldwide keep in touch, work more productively and relay vital information during emergencies with cellular phones and computers manufactured with beryllium-containing materials.

Connecting continents: On the ocean floor, copper beryllium housings protect the electronics that allow fiber optic cables to function flawlessly. These barrel-like housings, which resist highly corrosive sea water and extreme pressures, show little deterioration even after decades of service.

Enabling multi-function devices: On the ground and through the air, society stays connected thanks to beryllium and beryllium-containing materials:

- Battery contacts and electronic connectors in cell phones and portable electronics are made with copper beryllium alloys. The material meets stringent requirements for durability, weight savings, electrical conductivity and corrosion resistance in all extremes of weather and temperature.
- Copper beryllium alloys provide "spring memory" that ensures continuous, fatigue-free electrical connections through constant use, openings and closings, vibrations and accidental drops.
- The low electrical resistance and high thermal conductivity of beryllium-containing materials supports the convergence and miniaturization of multiple functions – phone, camera, MP3 player – into a single, lightweight, compact device.

Advancing digital technology: High-performance processors pack more and denser layers of high-frequency circuits into smaller packages. That means higher processing speeds and better performance for personal computers, routers, and the internet, as well as radars, avionics, and defense systems. The exceptional thermal conductivity and insulating properties of beryllia ceramic protect these systems from the potentially crippling effects of intense heat generation. With thermal conductivity up to 10 times greater than that of alumina ceramic, beryllia ceramics remain the insulators of choice for such high-frequency circuits.

Medicine

Advances in imaging equipment, diagnostics and laser medicine have been enabled by the strength and stability of this versatile metal.

The special properties of beryllium are essential to medical technologies that save and enhance lives.

Improving imaging: Because it is strong, stable, can handle elevated levels of heat resistance and is highly transparent to x-rays, beryllium, in thin foil form, has long been critical to the operation of medical and scientific x-ray equipment. Beryllium foil provides the window through which tissuepenetrating x-rays are focused, while maintaining the vacuum inside the x-ray tube generator.

Beryllium foil remains indispensible for high-resolution medical radiography, including CT scanning and mammography. Beryllium in newer generation mammography equipment enables a lower radiation dose scan with significantly finer tumor resolution, enabling breast cancer detection at its early, most treatable stages.

As imaging technology advances, beryllium continues to meet the need for x-ray tube windows with greater strength, stability, heat resistance and xray transparency.

Miniaturizing optical lasers: Medical lasers made with beryllia ceramics help ophthalmologists to restore or improve eyesight for millions. Beryllia ceramic is the only material that offers the thermal conductivity, strength and dielectric properties required to contain and control these tiny, high- powered gas laser bores.

Simplifying surgery: Copper beryllium connectors transmit precise electrical signals to delicate surgical instruments and monitoring devices used in the newest, non-invasive surgical techniques. Such techniques reduce patient trauma and infection risk, while speeding the process of healing and recovery.

Analyzing blood: Beryllium is also used in components of the analytical equipment used to analyze blood for HIV and other diseases, offering the precision and reliability that doctors and patients demand.

Transportation

Used in automobile airbags and electronic braking systems, weather forecasting satellites, chemical detection, fire suppression sprinkler systems and emergency rescue equipment, beryllium helps to keep us safe.

Ready for use in a split second: Beryllium's capabilities are used in airbags, fire sprinkler systems, aircraft bushings and bearings, power steering and electronic control systems, anti-lock brakes, undersea earthquake and tsunami detection monitors, air traffic control radar and weather forecasting satellites. Beryllium's strength, fatigue resistance and corrosion resistance can help to save lives and protect property in a fraction of a second.

Reliable for decades: Critical washers, made of nickel beryllium alloys, enable overhead sprinkler heads to deliver full water pressure in the event of fire, while maintaining a leak-tight seal for years of inactivity. Such reliability ensures that thousands of sprinkler systems, installed over many decades, continue to stand by.

Spark-free: The non-sparking and non-magnetic properties of beryllium help protect oil and gas exploration crews (and other people working around combustible gases) from risk of accidental explosions from sparks generated by tooling and equipment.

Ensuring child safety: Beryllium x-ray windows in hand-held analytical devices allow retailers and day care centers to ensure toys are free of lead and other harmful heavy metals.

Breathing easy: Beryllium alloy components ensure the reliability of pressure controls that deliver air from lightweight breathing systems used by firefighters and rescue crews.

Discovery

From heat shields that protected NASA's Mercury astronauts, to the orbital telescopes of tomorrow, beryllium materials have had a front seat in our nation's extraordinary exploration of space.

Beryllium Maintains a Legendary Role in Constant Pursuit of Knowledge

From NASA's earliest days, when beryllium heat shields protected Mercury spacecraft during re-entry, scientists, designers, and engineers continue to depend on this stiff, lightweight and versatile material to meet their most demanding challenges.

Orbiting the earth: Beryllium serves on current NASA vehicles including the Space Shuttle, where it adds strength, dissipates heat and lightens weight in window frames and door systems. Beryllium components also fly in the Spitzer Space Telescope.

Roving the Red Planet: Two Mars Rover vehicles, Spirit and Opportunity, have far exceeded original expectations. Aluminum beryllium components helped protect the rovers on their landings, and then served again to unfold their drive-off ramps. Aluminum beryllium parts used in the Rovers' rock exploration tools have helped our understanding of the planet.

Fixing Hubble: When the Hubble space telescope could not see clearly, its new "corrective lenses" were mounted in beryllium fixtures that met the requirements for lower weight, high stiffness and resistance to dimensional distortions brought on by extreme temperatures.

Looking ahead: The next-generation James Webb Space Telescope, scheduled to deploy in 2014, will depend on a 6.5 meter beryllium mirror to see objects 200 times fainter than visible before. Such mirrors must combine high stiffness and lightweight with an extraordinarily smooth, precise and defect-free surface. And, they must retain their visual quality for decades in deep space, where temperatures never exceed minus 253 degrees Centigrade.

Recreating the conditions after the "Big Bang". Scientists are utilizing beryllium components in earth-based particle accelerators to ensure high energy collisions of subatomic particles, recreating the conditions that could provide clues as to how the universe was formed. In the European Organization for Nuclear Research's (CERN) $10 billion Large Hadron Collider underground near Geneva, Switzerland, beryllium beam pipes surround the collision regions where experiments will be conducted.

Oil, Gas and Alternative Energy

Beryllium is in wide use in the energy field to extract oil and gas, and has a big role in helping to find tomorrow's clean and affordable energy sources.

Beryllium and beryllium-containing materials lower the cost of oil and gas exploration on land and in deep water. They're also helping lead the way to new and affordable sources of traditional fossil-based fuels, and increasingly, to exciting alternative energy sources.

Helping Stop the Leaking during the Oil Spills

Copper beryllium materials played an indispensible role in capping the Macondo Well in the Gulf of Mexico oil spill disaster.

Due to the material's incredible strength, resilience to stress and ability to slip over other materials without galling, copper beryllium clamps were used to securely fasten the capping stack to the damaged well pipe.

In addition, only copper beryllium has the strength, flexibility and transparency to magnetic fields that allowed the sensitive equipment to locate the blown out well and determine the precise location of the drill bit for the relief well bore.

Supporting Oil Exploration and Production

Alloys containing beryllium are used to extend the reach and effectiveness of oil and gas drilling equipment. Exploration crews depend on beryllium's non-sparking properties to reduce explosion and fire risks.

New directional drilling techniques, developed in the late 1990s, are helping rig crews reach more remote deposits. Yet, they depend on material that can operate reliably with electrical controls far from the drill head under high stress and elevated temperatures to prevent very costly drilling failure. Non-sparking copper beryllium alloy materials, which combine strength, conductivity, durability, and corrosion resistance, help make this possible.

Beryllium-containing x-ray tubes and detectors monitor the well flow rates of water and oil and gas as these materials are brought to the surface.

Capturing Solar's Potential

Alloys containing beryllium provide the thermal management, conductivity, and strength required in the electrical terminals that join the components of thin-film solar panels.

The superior thermal properties of beryllia ceramic allow concentrator photovoltaic (CPV) solar cells to operate at very high solar concentrations — 1,000 times the intensity of the sun— yet keep vulnerable electronics in the CPV cool enough to operate efficiently. This technology is already delivering electricity to power grids in areas throughout the world with abundant sunshine.

Containing Nuclear Fusion

Fusion reactors employ beryllium for its neutron reflecting and moderating properties, as well as its ability to withstand the extraordinarily high temperatures associated with fusion energy. Fusion holds promise for delivering virtually unlimited sources of energy with virtually no greenhouse gases and little nuclear waste compared to traditional nuclear fission reactors.

Beryllium has been qualified for use in the International Thermonuclear Experimental Reactor (ITER) project, a partnership that will operate the world's largest prototype fusion reactor. Beryllium components are already proven in the smaller Joint European Torus (JET) experimental reactor located in England.

Defense and Security

The armed forces of the world rely on beryllium – a critical component of weapons, guidance, surveillance and reconnaissance systems.

Beryllium is crucial to the defense of the nation, the protection of our allies and the security of the homeland. The U.S. Department of Defense (DoD) reported in 2008 that of all the metals used in its systems, only high purity beryllium was deemed "critical." DoD stated that beryllium is "essential for important defense systems and unique in the function it performs." NATO and the EU have presented similar conclusions.

A major application for copper beryllium alloys in aerospace is as the female connector terminals in the thousands of electronic and electrical connectors used in every aircraft, which are subject to intense vibration and stresses, yet must perform reliably over the 30 – 40 year lifespan of an aircraft.

Systems. Military systems depend heavily on electronics for navigation, target acquisition and firing. In critical situations and equipment, stiff, lightweight beryllium components ensure precise operation under extreme conditions.

In military fighter jets, pure beryllium saves weight critical to speed and maneuverability, while also ensuring razor-sharp targeting and strike capabilities. Copper beryllium is used for electrical connectors, fasteners and structural components in fixed-wing aircraft and fighters including the:

- F-35 Lightning II Joint Strike Fighter.
- F-22 Raptor.
- F-18 Super Hornet.
- F-16 Fighting Falcon.

- F-15 Strike Eagle.
- Eurofighter.
- BAE Tornado.
- Dassualt Rafael.

In optical systems of military helicopters, beryllium components are designed into enhanced surveillance and targeting systems that help keep crews safe.

The nation's unmanned aerial systems count on beryllium optical systems for real-time imagery and targeting on surveillance and reconnaissance flights.

For battle tanks on the move, stiff beryllium mirrors dampen vibration and provide a jitter-free optical path for targeting and firing controls.

Beryllium is also integral to the airborne equipment used to detect and destroy improvised explosive devices (IED) and tactical mines.

In emerging guided missile defense systems, beryllium is critical to assure a first line of defense in directing, targeting and ultimately destroying missile threats.

U.S. military satellites rely on beryllium metal for structural and dimensional stability, as well as reliability, in the electrical systems that deliver reliable intelligence from space.

Command and Control Communications: Military communications depend on copper beryllium alloys in network hubs, switches and routers. The strength, electrical and thermal conductivity of this material ensures reliability while maximizing signal speed and bandwidth.

Homeland Security: Behind the scenes at airports, ports, border stations and other public assets, beryllium and beryllium-containing materials support surveillance, inspection and countermeasures vital to security. At countless locations, beryllium components operate in the x-ray machines.

Transportation

Beryllium alloys are used in automobile components and airplane equipment to ensure the reliable operation of vital equipment and to enhance fuel efficiency. Beryllium helps get us from here to there.

Benefiting drivers everywhere: Today's cars and trucks include technologies and innovations that make driving safer, as well as more comfortable and fuel efficient.

Lightweight beryllium alloy connectors are used throughout the electrical systems of cars and trucks for their reliability and to improve vehicle fuel efficiency. Copper beryllium components are found in traction controls, transmissions, electric motors, anti-lock braking and fuel injection systems. Further, copper beryllium electrical connectors are used in electrically assisted steering systems that are replacing older, heavier hydraulic and electromechanical systems.

In commercial aircraft, landing gear bushings and bearings made from copper beryllium handle great compressive and wear forces despite corrosive atmospheres and exposure to wide ranges of temperature. The higher strength of copper beryllium compared to alternative bronze landing gear materials allows the bearings to be made smaller and lighter. These weight savings, in turn, make planes more fuel efficient which results in reduced exhaust emissions.

Beryllium is a crucial component of the electronic safety and navigation technology, including Global Positioning Systems (GPS) used for air-traffic control and air route surveillance.

Beryllium Alloys

Dilute alloys of base metals that contain a few percent of beryllium in a precipitation-hardening system are the principal useful beryllium alloys manufactured today. Although beryllium has some solid solubility in copper, silver, gold, nickel, colbalt, platinum, palladium, and iron and forms precipitation-hardening alloys with these metals, the copper-beryllium system and, to a considerably lesser degree, the nickel-beryllium alloys are the only ones used commercially.

Other than the precipitation-hardening systems, small amounts of beryllium are used in alloys of the dispersion type wherein there is little solid solubility (aluminum and magnesium). Various amounts of beryllium combine with most elements to form intermetallic compounds. Development of beryllium-rich alloys have been chiefly confined to the ductile matrix beryllium-aluminum, beryllium-copper solid solution alloy with up to 4% copper, and dispersed-phase type alloys having relatively small amounts of compounds (0.25 to 6%), chiefly as beryllium oxide or intermetallics, for dimensional stability, elevated temperature strength, and elastic limit control.

Beryllium-copper Alloys

Commercial alloys of the beryllium-copper group are divided into cast or wrought types, usually in ternary combination such as copper-beryllium-cobalt. Alloys with higher beryllium content have high strength while those with low beryllium have high electrical and thermal conductivity.

Age-hardenable copper-beryllium-cobalt alloys offer a wide range of properties, because they are extremely ductile and workable in the annealed condition and are strong and hard after precipitation or aging treatment. Cobalt is added to inhibit grain growth and provide uniform heat-treatment response. These alloys also have inherent characteristics of substantial electrical and thermal conductivity and resistance to corrosion and wear, being protected by beryllium oxide films which impart this property to all materials containing beryllium. These age-hard enable alloys resist dimensional change and fatigue.

Alloys with Nickel and Iron

Nickel containing 2% beryllium can be heat-treated to develop a tensile strength of 1700 MPa with 3 to 4% elongation. Little commercial use is made of the hard nickel alloys although they have been employed, principally as precision castings, for aircraft fuel pumps, surgical instruments, and matrices in diamond drill bits. Another nickel alloy with 2.0 to 2.3% beryllium, 0.5 to 0.75% carbon, and the balance of nickel and refractory metals has been used for mold components and forming tools for glass pressware of optical and container quality. Thermal conductivity, wear resistance, and strength, coupled with unusual machina-bility for a nickel-beryllium alloy, make this alloy particularly advantageous for glassware tooling.

Wrought beryllium-nickel contains about 2% beryllium, 0.5% titanium, and the balance nickel. Casting alloys contain a bit more beryllium (2 to 3%) and, in one alloy, 0.4% carbon. Arsenic in the case of beryllium-copper alloys, mechanical properties vary widely, depending on temper condition — from 310 to 1586 MPa in tensile yield strength and Rb 70 to Rc 55 in hardness at room temperature. The alloys retain considerable yield strength at high temperature: 896 to 1172 MPa at 538 °C. They also have good corrosion resistance in general atmospheres and reducing media. Because beryllium is toxic, special precautions are required in many fabricating operations. The wrought alloy is used for springs, bellows, electrical contacts, and feather valves, and the casting alloys for molding plastics and glass, pump parts, seal plates, and metal-forming tools.

Attempts have been made to add beryllium to a number of ferrous alloys. Small amounts are used to refine grains and deoxidize ferritic steels in Japan, and promising properties have been developed for austenitic and martensitic steels. Stainless steels (iron-nickel-chromium) may be made maraging by adding 0.15 to 0.9% beryllium, developing strengths as high as 1800 MPa as cast or 2300 MPa as rolled while retaining their oxidation- and corrosion-resistant characteristics. Amounts of 0.04 to 1.5% beryllium have been added to various iron alloys containing nickel, chromium, cobalt, molybdenum (Mo), and tungsten for special applications such as watch springs.

Beryllium-base Alloys

Three types of beryllium-base alloys are of interest. These consist of dispersed-phase types containing up to 4% beryllium oxide; ductile-phase or duplex alloys of beryllium and aluminum, particularly 38% aluminum in beryllium; and solid solution alloys of up to 4% copper in beryllium.

Dispersed-phase alloys containing oxides, carbides, nitrides, borides, and intermetallic compounds in a beryllium matrix are chiefly characterized by increased strength and resistance to creep at elevated temperatures. Major commercial alloys in this series are of the fine-grain, high-beryllium oxide (4.25 to 6%), hot-pressed types such as materials used for inertial guidance instruments characterized by high dimensional stability, high-precision elastic limit (55 to 103 MPa), and good machinability.

The 62% beryllium, 38% aluminum alloy under aluminum alloys was developed as a structural aerospace alloy to combine high modulus and low density with the machining characteristics of the more common magnesium-base alloys. This alloy in sheet form has at room temperature about 344 MPa ultimate strength, 324 MPa yield strength, and about 8% elongation. It is also produced as extrusions. It has a duplex-type microstructure characterized by a semicontinuous aluminum phase. Other alloys of the beryllium-aluminum type have been reported, but the 62% beryllium, 38% aluminum alloy is the most used.

Intermetallic Compounds

Beryllium, combined with most other elements, forms intermetallic compounds with high strength at high temperature (up to 552 MPa modulus of rupture at 1260 °C), good thermal conductivity, high specific heat, and good oxidation resistance. The beryllium oxide film formed by surface oxidation is protective to volatile oxides (molybdenum) and to elements of high reactivity (zirconium and titanium) at temperatures 1100 to 1500 °C and for short times up to 1650 °C.

The beryllides are of interest to the nuclear field, to power generation, and to aerospace applications. Evaluation of the intermetallics as refractory coatings, reactor hardware, fuel elements, turbine buckets, and high-temperature bearings has been carried out.

References

- Titanium: newworldencyclopedia.org, Retrieved 17 June, 2019
- Properties-and-applications-of-titanium-alloys-A-brief-review- 283863116: researchgate.net, Retrieved 04 June, 2019
- Heat-treatment-of-titanium-alloys- 90907: industrialheating.com, Retrieved 19 April, 2019
- Titanium-alloys-and-its-heat-treatment-metallurgy-26608: engineeringenotes.com, Retrieved 26 June, 2019
- Beryllium: newworldencyclopedia.org, Retrieved 31 March, 2019
- Beryllium-alloys, materialsparts-and-finishes: what-when-how.com, Retrieved 18 May, 2019

Permissions

All chapters in this book are published with permission under the Creative Commons Attribution Share Alike License or equivalent. Every chapter published in this book has been scrutinized by our experts. Their significance has been extensively debated. The topics covered herein carry significant information for a comprehensive understanding. They may even be implemented as practical applications or may be referred to as a beginning point for further studies.

We would like to thank the editorial team for lending their expertise to make the book truly unique. They have played a crucial role in the development of this book. Without their invaluable contributions this book wouldn't have been possible. They have made vital efforts to compile up to date information on the varied aspects of this subject to make this book a valuable addition to the collection of many professionals and students.

This book was conceptualized with the vision of imparting up-to-date and integrated information in this field. To ensure the same, a matchless editorial board was set up. Every individual on the board went through rigorous rounds of assessment to prove their worth. After which they invested a large part of their time researching and compiling the most relevant data for our readers.

The editorial board has been involved in producing this book since its inception. They have spent rigorous hours researching and exploring the diverse topics which have resulted in the successful publishing of this book. They have passed on their knowledge of decades through this book. To expedite this challenging task, the publisher supported the team at every step. A small team of assistant editors was also appointed to further simplify the editing procedure and attain best results for the readers.

Apart from the editorial board, the designing team has also invested a significant amount of their time in understanding the subject and creating the most relevant covers. They scrutinized every image to scout for the most suitable representation of the subject and create an appropriate cover for the book.

The publishing team has been an ardent support to the editorial, designing and production team. Their endless efforts to recruit the best for this project, has resulted in the accomplishment of this book. They are a veteran in the field of academics and their pool of knowledge is as vast as their experience in printing. Their expertise and guidance has proved useful at every step. Their uncompromising quality standards have made this book an exceptional effort. Their encouragement from time to time has been an inspiration for everyone.

The publisher and the editorial board hope that this book will prove to be a valuable piece of knowledge for students, practitioners and scholars across the globe.

Index

P

R

S

T

V

W

Y

www.ingramcontent.com/pod-product-compliance
Lightning Source LLC
LaVergne TN
LVHW080648210926
848485LV00003B/346